Finish Carpenter's Manual

by Jim Tolpin

Craftsman Book Company
6058 Corte del Cedro / P.O. Box 6500 / Carlsbad, CA / 92018

Acknowledgements

I would like to thank the following people for sharing some of their knowledge and skills with me in the course of writing this book.

Eric Almquist Sebastian Eggert Richard Hall Raoul Lazano
John Christensen Jim & Christine Gurman Cliff Jones Mike Parent
Pat Cudahy David Hageman David Kitchen John Penczer

I am especially indebted to *Dennis Callesen* and *Ken Kellman*, finish carpenters *par excellence*, who reviewed and made invaluable contributions to this book.

I dedicate this book to the memory of my grandfather: Samuel Rosoff

Some photographs, tools and equipment shown were furnished by the following manufacturers and suppliers. Full addresses and phone numbers are listed on page 205.

American Design & Engineering, Inc. Miter saw stand
Bosch Power Tool Corporation Belt sander with frame attachment and random orbit sander
By George Enterprises, Inc. Baluster installation jig
Eldenwood Enterprises Canvas bag
GIL-LIFT Cabinet dolly and installation lift
Hitachi Power Tools Compound miter saw and combination jointer-planer
Iron Age Safety Shoes Steel toed running shoes
Journeyman Products, Ltd. Fastener carrying bucket
KNAACK Manufacturing Co. Tool storage box
Maizefield Mantels Co. Mantels & staircases
Makita U.S.A., Inc. Palm-size random orbit sander, cordless panel saw and hammer-drill type screw driver
Matrix Enterprises, Inc. Saw protractor
Occidental Leather Finish carpenter's tool belt
Penzotti Pete's Knee pad pants
Price Brothers Tools Water level
Rousseau Co. Table saw extension table
Senco Products Pneumatic nailer
Wedge Innovations Electronic jamb level
Wood Dynamics Table saw throat plate
Woodstock International, Inc. Edging round-over plane
Woodworker's Supply of New Mexico Push blocks

Library of Congress Cataloging-in-Publication Data

Tolpin, Jim, 1947-
 Finish carpenter's manual / by Jim Tolpin
  p. cm.
 Includes index.
 ISBN 0-934041-82-2
 1. Finish carpentry—Handbooks, manuals, etc. 2. Finish
carpentry—Estimates—Handbooks, manuals, etc. I. Title
TH5640.T65 1993
694'.6—dc20   93-15099
          CIP

Contents

Chapter 1

Modern Finish Carpentry

Finish carpenters today do some of the same work their grandfathers and great-grandfathers did 50 or 100 years ago. But construction changed dramatically during the 20th century. Techniques changed. Materials changed. The trades became more specialized. And some types of work simply vanished. For example, how long has it been since you carved the gooseneck for a handrail, built cabinets on site, or milled window sash?

Some tasks that were once considered finish carpentry aren't any more. Fitting exterior siding and trim were once finish carpentry. So was hanging windows and exterior doors. Now these jobs are usually done by framing contractors. And other jobs, such as hanging drywall or installing suspended ceilings, are now primarily done by specialty subcontractors.

Even so, there's plenty of work left for a good finish carpenter to do, and a good living to be made by doing it! It's important, however, to master (or at least have a thorough knowledge of) all the aspects of the trade, past and present. Even if you don't hang windows, you still have to know the difference between a window that's hung right and one that isn't. It's senseless to spend an hour carefully trimming out the interior of a window — and then have it yanked because the sash is misaligned. An old-timer who's mastered all the trade wouldn't make that mistake. And you shouldn't, either.

That's why I'll cover all the basics of finish carpentry in this manual. Some of this you'll use fairly seldom. But every professional finish carpenter (and that's you!) is expected to know it.

Finish Carpentry Skills

Finish carpenters need all the skills (and most of the tools) of a framing carpenter — and more besides. But, obviously, the trades are very different. For example, framers work in two dimensions — length and width. Cut it, stand it up, and nail it in place! Precise fitting is seldom practiced or needed.

On the other hand, we finish carpenters work in all three dimensions — length, width, and depth. Most of what we do requires precision fitting and careful installation.

A framer's work will be covered and forgotten long before a building is occupied. But what we do will be on display for the entire life of the building. The smooth, even surfaces we install have to fit right because they're always in full view. That means measuring and cutting must be precise. We have to calculate and cut compound angles exactly. We have to fit smooth trim pieces on rough textured surfaces that are neither plumb nor square. It takes skill to do that, skill to create a gap-free joint, skill to shape, smooth and fit wood until it's perfect. The finish carpenter carries a big burden. Dennis Calleson, a master of the trade, puts it this way: "A good finish carpenter is the conscience of the construction project."

Precise installation is the essence of finish carpentry. If you enjoy joining smooth, sleek surfaces to create a perfect match where materials meet, congratulations! You have the makings of a true professional craftsman. You'll have no trouble absorbing (and applying) all the information between the covers of this book.

Scope of Modern Finish Carpentry

Even though perfect installation is the essence of finish carpentry, finish carpenters need more than installation skills. You'll find there are many times when the architect's plans and details won't show finish work details. It will be up to you to assume responsibility for layouts and joinery choices. You'll also have to write up bills of materials, develop cut lists, and on some jobs you may even have to design the trim elements from the ground up.

Here's a list of what I consider to be the scope of the modern carpenter's trade. It's also what I intend to cover in this manual.

1) Material take-off and layout of finish elements.

2) Preparation of interior surfaces to receive finish work.

3) Hanging and trim out of interior doors.

4) Trim out of interior window casings and jamb extensions.

5) Installation of running moldings such as base, cornice, chair and picture rails.

6) Installation of wainscoting, and other types of wall paneling.

7) Installation of wood strip or paneled ceilings, as well as applied beams, posts and corbels.

8) Fabrication and installation of closet shelving and clothing poles.

9) Installation of wood flooring.

10) Installation of factory-made cabinetry.

11) Fabrication and installation of fireplace surrounds and mantels.

12) Installation of interior stairways and balustrades.

13) Installation and trim out of entry ways.

14) Installation of door and bathroom hardware.

Since almost nothing gets done in finish carpentry without the right tools, I'm going to begin by describing the tools you'll need. That's the topic for Chapter 2.

Finish Carpentry Hand Tools

It's hard to imagine a trade that requires more tools than finish carpentry. But then, finish carpenters have to work at close tolerances with a variety of materials, such as solid wood, plywood, metal and plastic.

If you're like me, you take real pride in your tool collection. We finish carpenters tend to lust after quality hand tools. If nothing else, being in this trade gives us a fine excuse to buy and own a really splendid set of tools.

The basic clothing and safety equipment needed by a professional finish carpenter include the following:

Workpants with built-in knee pads. I don't like strap-on knee pads. They cut off circulation and are uncomfortable to wear for any length of time.

Workboots or shoes. Running shoes are comfortable, support your arches, and are cool in the summer — but they don't do much to protect your toes! Recently, however, steel-toed running shoes have come onto the market. See Figure 2-1. These offer you a safe choice between a lighter-weight shoe and the standard steel-toed leather workboots.

Hearing and eyesight protection. Foam ear plugs are comfortable and can reduce hearing loss when you work with noisy equipment. Be sure to wear impact-rated goggles or side-shielded tempered safety glasses when nailing or operating any power equipment. Buy comfortable eyewear and make it a habit to wear it throughout the work day.

Lung protection. Wear a disposable, two-headstrap, double-thickness paper mask during sanding and cutting operations. Use a carbon-filtered respirator when toxic fumes are present during finishing processes.

First aid kit. Carry a selection of adhesive bandages, some gauze pads and tape, scissors, antiseptic, and splinter-tweezers in a dust-free box.

Soap and shop towels. Stock a grease-removing hand soap so you don't get tool oil on the raw finish stock.

Floor sweep and pan. Sweep up those stock cutoffs so they don't get underfoot and break your bones.

Figure 2-2 shows the author, geared up and ready for work.

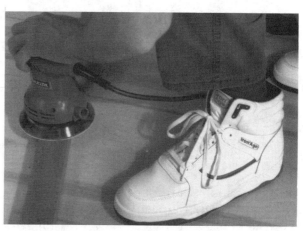

Shoes courtesy: Iron Age Safety Shoes

Figure 2-1　**Steel-toed running shoes**

Photo by Pat Cudahy

Figure 2-2   **Finish carpenter at work**

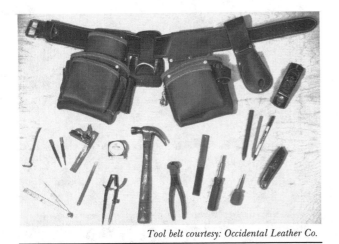

Tool belt courtesy: Occidental Leather Co.

Figure 2-3   **Tool belt with tools**

The Tool Belt

A good tool belt is your constant companion. It's like your right (and left!) hand man. The belt keeps all the tools you need for a wide range of tasks readily at hand. Figure 2-3 shows my belt laid out with the tools I typically carry.

A wide leather belt is essential to transfer the weight of all those tools and fasteners to your hip. Make sure it's comfortable and feels right. A sheepskin liner adds padding for comfort and reduces heat chafing in hot weather.

Leather pouch tool belts are traditional. Some newer belts have pockets made of synthetics. These belts are lighter and may possibly last longer, but I still like the feel of a solid leather belt. It's stiff enough to maintain pocket shape but still flexible enough to bend in tight situations. If the weight is a problem, you can add a pair of suspenders to transfer a portion of the load off your back and onto your shoulders.

Belts designed for finish work have numerous small pockets within the larger pouches. These pockets provide homes for the smaller tools a finish carpenter must carry.

Use the pouches without the pockets for nails and screws. Keep your belt organized. Tools used in your primary hand, such as pencils, knives, block planes, go to that side. Secondary tools, such as nail sets, fasteners and combination squares, go on the other side of the belt.

Having put in my plug for tool belts, it's only fair that I mention the classic carpenter's work overalls. Some finish carpenters can make do with just their overall pockets, eliminating the need for a tool belt altogether. Not me. I apprentice-trained under a tool-belt man and don't feel comfortable with just overalls. But I'll give equal time to the opposition.

For starters, a work overall man will tell you that overalls protect your clothes. Then he'll go on to say that the extra layer of cloth makes knee pads unnecessary. Plus, without a tool belt, you can work in tighter places. And working without a tool belt reduces damage to wall surfaces — all the fault of the tool belt guys, of course!

Also, small tools like nail sets and screw bits are easier to find in overalls (that's the theory anyway). And of course, the bib has pockets for pencils, 6-inch steel rulers, etc. And last, but not least, I'll admit to all the overall guys out there, when you're wearing overalls there's always a rag handy!

Here's a list of the tools typically found in the finish carpenter's belt — or overalls.

- Tape measure, 1 inch x 25 feet or ¾ inch x 12 feet
- 6-inch steel ruler
- 6- or 12-inch combination square
- Number 2½ pencils
- Marking scribe or awl
- 13-ounce straight-claw hammer
- Nail set
- Nail punch
- Utility knife
- Mini Wonderbar® or painter's pry bar
- Block plane
- Four-in-hand file/rasp
- Bevel gauge
- Exchangeable bit screw driver set
- End snips or diagonal cutters
- Compass scribe
- 4-ounce glue bottle with cap

Additional Hand Tools

Of course, you're going to need more tools than will fit comfortably on a tool belt. The hand tools on the list that follows should meet the hand-work needs for most finish carpentry tasks.

Saws

- Japanese-style combination saw — combines rip and crosscut functions
- Compass saw
- Coping saw with 20-teeth-per-inch blade. Use primarily for coped molding joints.
- Dovetail saw
- Flush-cutting saw. Use to cut bungs, pins, and other protrusions flush to a surface.
- Drywall saw

Measuring Devices

- Tape measure, 100 foot — for layout and material estimates
- 2-foot framing square
- 4-foot panel square
- Angle finder and angle divider
- Gooseneck dividers, with the pencil on the side opposite the protrusion of the swing arm
- Trammel points — for laying out arched moldings of any radius
- Pinch sticks. Use to record inside measurements; one set to measure 22 to 38 inches and one set to measure 36 to 68 inches. The ends of the sticks should be cut at a 45-degree angle.
- Plumb bob
- Chalk line
- Straightedges, several sizes. Use to check straightness and extend bubble levels to specific lengths. Use a 60-inch straightedge to check a closet header, for example.

Levels

- Builder's transit level — for layout of horizontal trim elements. This is an optional tool, but it's a great time-saver.
- Torpedo level
- Rail levels, four are nice: 24-inch, 30-inch, 48-inch, and a 78-inch jamb level
- Electronic level — to take pitch and angle readings and to accurately calibrate bubble levels

Planes

- Low-angle block
- Jack
- Rabbet
- Mortise
- Edger — for breaking sharp edges

Chisels

- Set of butt chisels to 2 inches
- Push chisel, 1¼ inch — for trimming to fit

- Crane neck — for trimming off bungs and pegs
- Corner chisel — for squaring butt mortises
- Set of palm carving chisels — for fine shaping or carving complex molding junctures

Shaping Tools

- Surform®
- Rat-tail file
- Set of 6-inch round and triangular files — for fine shaping work
- Set of cabinet scrapers: hand-held rectangle, gooseneck, and handle mounted
- Assortment of contoured sanding blocks

Holding Tools

- Carpenter's vise (clamps on sawhorses)
- Workmate® portable bench
- Quick-grip® clamps, a 12-inch pair is usually enough
- C-clamps, pairs in sizes from tiny to 6 inches. The more the merrier.
- Miter clamp — to hold mitered stock together while fastening
- Spring clamps, at least one pair

Drill Bits

- Standard jobber set to ⅜ inch
- Brad point set to ½ inch
- Vix® bits for #6 and #8 screws. Use to center pilot hole in hinge-screw installations.
- Speed-bore® or spade bit set to 1½ inches
- Hole saws: 1-inch, 1½-inch, 2½-inch and 3-inches for plumbing holes
- Screw piloting bits with countersink: #6 and #8 with ⅜-inch countersink and #10 with ½-inch countersink
- Plug cutters to produce ⅜- and ½-inch bungs to insert in countersinks; requires drill press or jig with hand drill

Glue and Caulk Applicators

- Refillable ½-pint plastic bottle with pointed tip
- Bottle with slotted tip and wet storage box for spline biscuit installations
- Glue injector — for remedial work on open joints or splits; a horse syringe works great
- Bottle with brush-headed applicator for contact cement
- Caulking gun. The best ones have an additional lengthwise stretcher. A cake icing tube also makes a good caulk applicator.

Mechanic's Tool Set

- Set of screwdrivers, include #1 and #2 square drivers
- Open-end wrenches
- Socket set
- Crescent wrench
- Slip-joint pliers
- Needle-nose pliers
- Vise grips
- Set of Allen wrenches

Rough Construction Tools

- Crowbar
- Wonderbar®
- 3-pound sledge
- Staple gun

Miscellaneous

- Moisture meter
- Calculator, preferably one that can add fractions
- Flashlight
- Electronic stud finder. Carry an extra battery to the field.

Sharpening Hand Edge Tools

Keeping your cutting tools sharp is an easy way to keep the quality of your work up and your stress level down. This means honing them back to razor sharpness again and again in the field. I use a set of waterstones and honing guides. See Figure 2-4

I have found that waterstones are better to work with than oilstones. Oilstones are messy. Even with care, the lubricating oil has a way of getting onto raw wood surfaces and wrecking the finish. They lose their flatness quickly. And since it's a real chore to reflatten them, they often stay out-of-flat. They're also slow to remove metal — perhaps twice as slow as a waterstone.

Waterstones are easy to keep flat. Edges honed on a waterstone aren't any sharper than those honed on an oilstone. But it's a lot less hassle to get a razor sharp edge, and keep it that way.

Sharpening with Waterstones

Here's the procedure for sharpening a blade with a waterstone:

1) Use three grits of stones in progression: 800 to 1200 to 6000.

2) Keep the stones immersed in water until you use them. While running the metal over the stone's surface, allow a slurry of water and grit to build up. Add only enough water to keep the slurry from drying and precipitating out the stone particles.

3) I recommend sharpening guides with rollers to hold the tool at a consistent angle. The angle depends on the type of tool being sharpened. In general, use 25 degrees for chisels and 30 degrees for plane irons.

4) Optional *micro bevels* can be worked into the blade to make honing faster. Lift the blade an additional 5 degrees on the stones. This creates another bevel about 1/32 inch wide. See Figure 2-5.

5) During the first part of the sharpening process, only the beveled side of the blade is honed. Remove the blade from the guide only after the bevel reaches a mirror polish on the 6000 grit stone. Turn it over and remove any burr on the back (flat) side. Do this by making several passes over the 6000 stone. See Figure 2-6.

Figure 2-4   **Honing guide in use on waterstone**

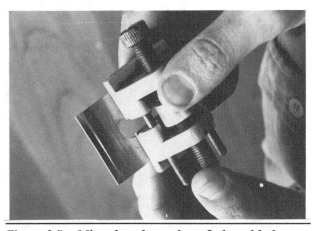

Figure 2-5   **Micro bevel on edge of plane blade**

6) You can keep the stones flat in the field by rubbing a coarser grade stone over a finer one. Be sure to add water for lubrication. Eventually, however, this creates a curve in the stone. To straighten the stone, rub it over a piece of ¼-inch plate glass or a steel flattening plate designed for this purpose. See Figure 2-7. Speed up the flattening process by dusting silicon carbide powder in the slurry. Use progressive grits from 60 to 120 to 220.

Wouldn't it be nice if we could make a living using just this collection of hand tools? Think of the peace and quiet of it all! But those days are history. To make a living at finish carpentry today, you have to own, and know how to use, a variety of power tools. And that's the subject of the next chapter.

Figure 2-6   **Removing burr on back of blade**

Figure 2-7   **Stone being rubbed on steel flattening plate**

Chapter 3

The Power Tools

This chapter describes the power tools commonly used by modern finish carpenters. I've included operating tips and safety cautions where they apply. You may not need to own each tool I describe, but I encourage you to read through the entire chapter to acquaint yourself with these specialized tools. I've divided them into two categories: hand power tools and stationary power tools.

Hand Power Tools

You'll use hand power tools to cut, drill, rout, join, plane, fasten and sand. Let's begin with the cutting tools.

Cutting Tools

Circular saws- A direct drive 7¼-inch saw is adequate for most cutting needs. If you intend to do a lot of paneling work, I recommend you also get a 3⅜-inch trim saw. Carbide blades make clean cuts over a long period of time so they're worth the extra money. Before using any circular saw, check the base for flatness and squareness to the blade at the "zero" setting. Figure 3-1 shows a 7¼-inch circular saw with an angle guide. Keep the saw, with its arbor wrenches and extra blades, in its own sturdy carrying box.

To make a circular saw more accurate and versatile, use an angle guide for indexing and guiding angle and square cuts. You can also make a *cross-cut guide* for trimming sheet stock and doors. See Figure 3-2.

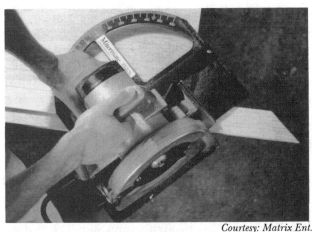

Courtesy: Matrix Ent.

Figure 3-1   **Using a circular saw with an angle guide**

½" hardwood plywood

¾" hardwood

50" or 38"

4"

X

X= Distance between edge of circular saw blade's table and the saw kerf

- 50" version for cross-cutting sheetstock
- 38" version for cross-cutting doors to height

Figure 3-2   **Cross-cut guide**

Courtesy: Hitachi, Inc.

Figure 3-3   Sliding compound miter saw

Courtesy: American Design and Engineering, Inc. (stand)

Figure 3-4   Miter saw on portable stand with a swing stop

Safety tip: Since you generally use a circular saw with one hand, supporting the stock with your free hand, keep aware of where your free hand is in relation to the travel of the blade. Make sure you support the stock against a solid surface when you make a cutoff. This helps keep the saw from binding and kicking back. An old carpenter once told me that each morning before he went to work, he looked at his hands and promised each of his fingers that he'd bring them home that night. To this day, keeping those fingers on his mind as he worked has kept them on his hand.

Jig saw- Spend the extra money and buy a good one. Quality jig saws cut smoothly and quickly with little vibration. The best models have variable speed controls and an adjustable reciprocating blade action. This feature produces fast cuts, even when ripping. Just like the circular saw, your jig saw deserves a box for blade changing tools and extra blades.

Chop saw- Today, at the heart of most finish carpenter's power tool collection, there's an 8½-inch blade-sliding compound miter saw. Figure 3-3 shows one. You can set up this new generation of chop saw to quickly cut an accurate compound angle across boards that are up to 12 inches wide.

For some tasks, such as cutting joints on wide cornice moldings, you can't beat this tool (See Chapter 11 on cornice installations).

The only disadvantage of the modern blade-sliding compound miter saw is that the cutting action is slower than the action of a standard chop saw. Because of this, some carpenters still prefer the chop saw for making quick, repetitive cuts in small molding stock.

A handy accessory to the compound miter saw — or chop saw — is a portable stand and stock support system. The best stands have an indexed sliding stop. See Figure 3-4.

For best results, use a carbide blade specifically designed for chop saws.

Safety tip: Don't remove the guard, and always wait for the blade to stop spinning before you extract cutoffs. An unpowered blade, even one that's moving slowly, can still take off a finger. I've seen it happen.

Reciprocating saw- This is really more of a rough construction tool than a finish tool. It's often called by the brand name *Sawzall*. It's handy for prep work, such as making alterations to framing. Carry a variety of blade lengths and include a metal cutting blade. See Figure 3-5.

Drilling Tools

Hand drills- The basic drill for a finish carpenter is a high quality ¼- or ⅜-inch chuck capacity, variable-speed, reversible, corded drill. Be sure to buy a good one. A quality drill lasts longer and

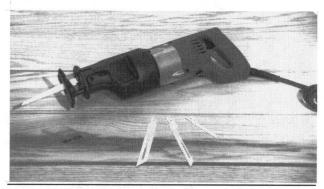

Figure 3-5 **A reciprocating saw with a variety of blades**

Courtesy: Makita, Inc.

Figure 3-6 **A hammer drill, useful for driving screws**

Figure 3-7 **One-horse plunge router with edge guide**

will have higher torque under load at low speeds. This means that when you use the drill to drive screws, it won't spin up to higher speeds and strip out the slot of the screw.

Include at least one cordless drill with a 9.6 volt battery in your collection. Always carry a spare battery and charger to the job site.

A hammer drill is handy for drilling anchors for attachments in concrete. You can also use it to drive screws. The hammer action helps keep the driver from stripping out the screw head. See Figure 3-6.

You'll also need a heavy-duty ½-inch drill for making large holes in wood.

Safety tip: If you have long hair, keep it tied back when you use any power tool. I've seen long hair get sucked into a drill's ventilation slot and wind its way around the armature. The drill sped right up to the carpenter's head with a vicious smack. He was also more than a little embarrassed walking into the barbershop with a ⅜-inch drill stuck on the side of his head!

Routers

Plunge router- A one-horsepower router with plunge capability will handle most of a finish carpenter's needs. See Figure 3-7. If you're going to rout lots of molding, I recommend a three-horsepower version with the capacity to accept ½-inch shank bits.

An adjustable fence-edge guide is handy for jobs where you have to space the cut in from the edge of a board.

Safety tip: Run the router along the work against the direction of the spinning bit. This prevents a "runaway" motion. Never make a cut so deep that it drags down the speed of the motor.

Trim router- A small router that you can use with one hand makes light work of edge shaping tasks. If you use it for laminate tasks, you'll want tilting and offset bases. See Figure 3-8.

Safety tip: Because you often use a trim router with just one hand, you should always be aware of where your free hand is.

Figure 3-8   **A trim router**

Figure 3-10   **An electric planer**

Figure 3-9   **Plate biscuit joiner in use**

Bits

A basic set of router bits for finish work includes:

1)   Round over bits in ⅛-, ¼-, and ⅜-inch radii

2)   Smaller pilot bearings for the above — to allow them to produce a stepped round-over

3)   ⅜-inch and ½-inch radius cove bits

4)   A ⅜-inch depth rabbeting bit. Use larger pilot bearings to reduce the depth of cut.

5)   Straight bits in various widths. Spiral cutting flutes make smoother cuts.

6)   Chamfer bit

7)   Hinge mortise bit. This is a straight-sided bit designed to cleanly cut a shallow mortise for hinge plates.

8)   Template following bits. Straight-sided bits with a bearing on the shank above the cutters. In use, the bearing rides against a template while the cutters mimic the pattern in the material below.

Unless you enjoy sharpening these tiny odd-sized bits of metal, use only carbide-tipped bits and send them out to a professional sharpener when they need sharpening.

Joining Tools

Plate biscuit joiner- This is a relatively new tool to finish carpentry, but it's one that's fast becoming a core addition to any collection. See Figure 3-9. As you'll see in later chapters, this tool replaces the older method of dowel joinery to align and strengthen butt joints. Splines are much faster to cut and install than dowels, and they hold better.

Shop-made layout sticks and support jigs are handy accessories for this tool. I'll discuss these more fully in Chapter 14.

Planer Tools

The electric planer- The small 3¼-inch wide planer works well for most tasks. Some versions of this tool have reversible carbide blades. This is the type I use and recommend. See Figure 3-10.

Electric hand planers take some getting used to. Practice first. It's real easy to take off more material than you mean to with this tool.

Safety tip: Take extra care when using this tool because of the exposed cutters. Don't set it down when it's still running without a holder designed for the purpose. This is another one-hand tool. Keep the other hand well out of the way.

Fastening Tools

Screw-shooting drills- These drills are designed to hold and drive drywall screws. If you'll be doing a lot of finish work with this type of screw, it's a useful drill. Generally though, a standard variable-speed drill, and especially a hammer drill with a magnetized driver bit, will serve as well.

Pneumatic finish nailers- Like the compound miter saw, the finish nail gun is finding its way into nearly every finish carpenter's tool kit. And for good reason too. It's usually better to shoot a nail than drive it in with a hammer. That way you'll avoid hammer-dimpling expensive finish materials and spreading oil from the nails to the wood. And split-outs are less common when driving nails near the edges and ends of boards.

Another advantage is that you can hold the stock in position with one hand and fasten it in place at the same time. But there are some disadvantages too. Air-driven nails sometimes don't hold as well. They won't suck a piece of molding to a wall deformity the way a hammered nail will. In addition, most air guns also need a compressor, hoses, and an adequate power source.

Most finish carpenters use at least two types of air nailers:

■ A *brad tacker* which shoots ⅝- to 1½-inch slight-headed nails. Some brands make two guns to span this size range.

■ A *finish nailer* which shoots 1¼- to 2½-inch finish nails.

An alternative choice is:

■ A *cordless, impulse-fired finish gun.* Some brands — *Paslode*, for example — can handle finish nails from ¾ inch to 2 inches. The advantage of this gun is that it doesn't require an air compressor and hoses for operation. The disadvantages are its size and weight. Also, there's the expense, and toxic off-gassing of the fuel cells. And it makes a louder pop when fired.

Safety tips: Treat pneumatic nailers with the respect you'd give a firearm. Make sure the gun is held securely against the stock before pulling the trigger. See Figure 3-11. Avoid "bounce nailing" trigger setups (the gun fires each time you push the safety bar at the nose of the gun). Never point the gun toward anyone, even if they are on the far side of a room. Keep your free hand out of the line of fire. An occasional double-fire can send a second nail flying through space. I have a hole in my hand to prove it.

Courtesy: Senco, Inc.

Figure 3-11 Pneumatic nail gun in use

Courtesy: Bosch, Inc.

Figure 3-12 **Random orbit sander with accessories**

Courtesy: Bosch, Inc.

Figure 3-14 **Belt sander with sanding frame**

Sanding Tools

Random orbit sander- These sanding machines are fairly new to finish woodworking. See Figure 3-12. They have several advantages over classic orbital sheet sanders. They remove material faster and leave a virtually mark-free surface. Some even remove dust as they sand — through suction holes in the sanding pad. You can also cover the pad with a buffer and use the tool as a polisher.

A real disadvantage to the random orbit sander is that the pads are circular. So, to sand into corners, you have to switch over to a standard sheet sander.

Sheet sanders- Most finish workers have two: the half-sheet version and the smaller, quarter-sheet sander. ("Sheet" refers to the standard size sheet of

sandpaper). Both do a good job of sanding surfaces relatively quickly, with a minimal amount of scratch marks. Here again, the more expensive machines do the best job. See Figure 3-13.

I recommend installing *Stikit* replacement pads on these sheet sanders. These pads let the machine take *Stikit* sandpaper, which comes in 100-foot rolls. This stuff is very durable, and best of all, you can change the paper in under ten seconds.

Belt sanders- A 3- x 21-inch belt sander is really all you need for most finish trim work. But you can sand out larger surfaces with a 4- x 24-inch model, and some carpenters like its additional weight. Other useful options include a dust bag, a belt cleaner, and a snap-on sanding frame which prevents side gouging during some types of sanding operations. See Figure 3-14.

Disc grinders- Small disc grinders with 4- or 4½-inch wheels can be very effective and versatile tools. Use them to quickly and cleanly remove metal, such as protruding screw heads, or to shape surfaces. With a sanding pad and disc installed, you can swiftly and accurately waste away wood to cut lines.

Figure 3-13 **Two sheet sanders (quarter-sheet and half-sheet)**

This is a tool you can do without. But once you own one, and learn to use it, you'll wonder how you ever made do without it.

Figure 3-15 **Builder's bench saw with extension tables and fence system**

Miscellaneous Power Tools

Shop vacuum- This tool is a must on the job site. Use it to clean up before and after installations, and to dust off stock before applying finishing materials.

Dremel grinder/sander- This isn't a "have-to-have" tool. But it's useful for sharpening all sorts of drill and router bits on the site. Some carpenters use it to make fine adjustments at the junctures of complex moldings (instead of a palm chisel).

The Stationary Power Tools

Now for the my second category of power tools — the stationary tools.

Builder's Bench Saw

Few finish carpenters lug their old 200-pound table saws to the job site any more. A new breed of lightweight, portable bench saws is now available for site work. With experience, and the right accessories, these saws can do nearly the same work as their hefty predecessors. Figure 3-15 shows a setup with an after-market extension table and fence system.

Take the time to carefully align the blades and rip fence to the miter gauge slots in the table surface. Adjust the blade-tilt stops for a true 90 and 45 degrees. Also, check the indicator on the rip fence to be sure the indicated measurement to the blade is correct.

A high-quality carbide combination blade is a must. Choose the thin-kerf variety made especially for these machines. Carry a cheap blade to use when cutting stock that might contain nails or glue lines.

Safety tip: Read your user's manual and follow the standard precautions for using a table saw. In addition, think about where you're going to set up your saw on the site. Stay away from high traffic areas. Make sure the floor is solid. Give yourself plenty of clearance to work on all sides.

Jointer/Planer

In addition to the builder's bench saws, a number of lightweight jointers and planers — and combination machines — are now available. They give you quality performance for a reasonable price. When set up securely, they can handle surprisingly large pieces of work.

Safety tip: Attach these machines securely to a pair of sawhorses. Always use pushblocks when passing boards over the cutter heads. See Figure 3-16.

Figure 3-16 **Combination jointer-planer with pushblocks in use**

Keep the table surfaces well waxed to prevent rust, reduce friction, and help prevent kickbacks.

Compressor

You need an air compressor to power most types of pneumatic nailers. Generally a ¾-horsepower model has enough capacity to supply two guns at once. A one-horsepower compressor is fine, but you'll run the risk of blowing the usual 20 amp, 120-volt circuit. You could use 220-volt power, but getting it is a real headache on some sites!

Buy good compressor hoses. Cheap hoses need constant mending and replacement. For less than 100-foot runs, ¼-inch hose is good enough. Go to ⅜-inch for longer runs to keep up enough air supply.

Make sure the fittings on the guns are compatible with the fittings on the hose. Carry adapters to the field in case you need to attach your hoses to somebody else's. Also, make sure you keep oil in the oil pan of the compressor and in the air lines to the guns. Read your user's manual and follow the suggestions of the manufacturer.

Chapter 4

Site Fixtures

I t's not enough to assemble a complete collection of finish carpentry tools. You also need a way to carry them to and around the construction site, and to store them out of harm's way. And you need fixtures to support the work and bring in light and power. You can either buy these accessories or make them yourself in your home workshop.

Tool and Fastener Storage

Site Tool Storage Boxes

A commercial-quality locking steel storage chest is essential if you have to leave your tools at the construction site overnight. They come in a wide range of sizes, and are both secure and fire resistant. The chest I use has casters so it can be moved around easily. See Figure 4-1. When I have to leave it overnight, I chain it to something permanently affixed to the site, like a metal soil pipe.

If you're willing to cart your tools home every night, you can build your own storage box out of wood. Figure 4-2 shows a versatile unit that includes both drawers and a work surface. Install a lockable steel bar through the work surface to overlap the door and drawer faces and keep them from opening during transportation.

Hand Tool Tote and Stool

I use a small tote box for hand tools that can also double as a step stool. The height is perfect for helping you reach most window and door head casings. It's a handy sidekick to the larger site boxes. Load it up with the tools you need for a specific task. A built-in carrying handle makes it easy to take from place to place. Figure 4-3 shows you my method for building a tool tote box.

Fastener Storage Systems

You'll use many types of fasteners on a construction site. In the past, carpenters designed and built their own tray boxes to handle a personal "filing" system. The commercial fastener totes available today, however, have

Photo courtesy: KNAACK Mfg. Co.

Figure 4-1   **Commercial tool storage box**

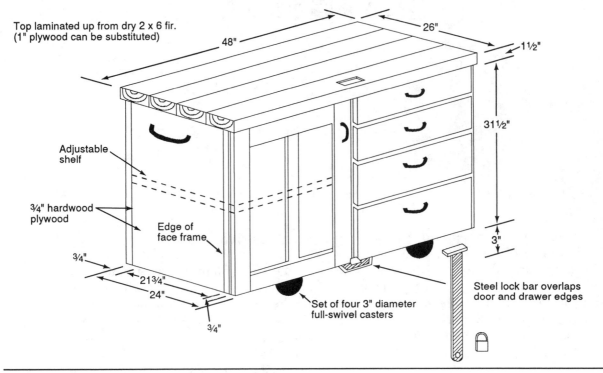

Figure 4-2   Shop-built tool storage unit with work surface

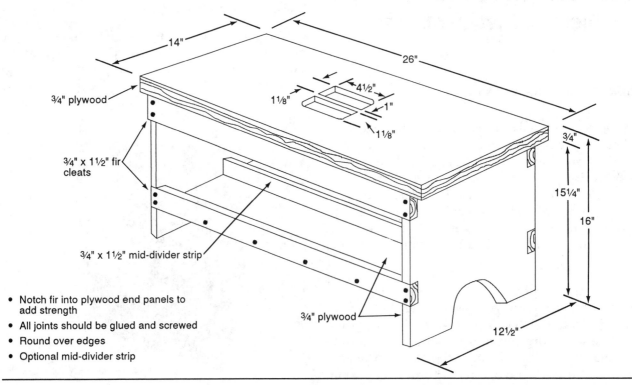

Figure 4-3   Shop-built tool tote and step stool

BAG-It courtesy: Eldenwood Enterprises
DROP-IN-THE-BUCKET courtesy: Journeyman Products, Ltd.

Figure 4-4 **Fastener carrying systems**

features that are hard to duplicate in the home shop. They organize your fasteners in a minimum amount of space and protect them from dust and debris.

The DROP-IN-THE-BUCKET™ system has five separate four-section inserts that nest neatly in a five-gallon bucket. BAG-It™ uses radially-divided canvas pouches with drawstrings. See Figure 4-4.

Getting the Power You Need

On new construction sites and on some remodels, getting power to where it's needed can be a real headache. Sometimes power is available only at the drop pole. A good solution to this problem is to run a heavy feed line to a secondary power feed box set up in the work area. From there, run short extension cords to where power is needed. Buy or build a box as shown in Figure 4-5. Unless you're also an electrician, have the wiring done (or at least checked) by a professional.

Because the box is going to get wet, use waterproof glue in the joints. It's also a good idea to paint the whole outside of the box.

Light Stands

Commercial light stands are extremely effective, rugged — and expensive. If your funds are limited, consider making a sturdy home-built stand. A knock-down base makes transportation

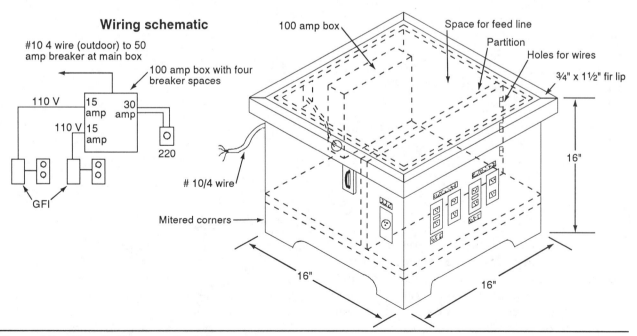

Figure 4-5 **Shop-built power feed box**

easy. See Figure 4-6. What you've got when you're done is a simple arm that will hold standard clip-on shop lights.

Work Supports

Sawhorses

Every carpenter has a favorite way of making up a pair of sturdy sawhorses. You can see mine in Figure 4-7. For a finish carpenter, light-duty horses are usually enough. These are made entirely of ¾-inch plywood. They're lightweight, a snap to assemble and they take up very little room in the back of a pickup. They may not look it, but they're surprisingly strong.

A bit of advice: Avoid using screws to attach fir 1 x 2s to the top edge of the girt. If you do, a saw blade is bound to find them!

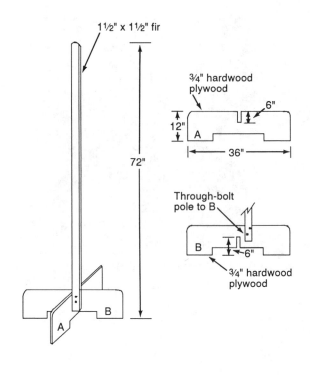

Figure 4-6   Shop-made light stand

Panel Lifts

When you're cutting drywall with a panel saw, a set-up that lifts the panel off the floor and supports the cut-offs is very handy. You can quickly and easily put together a low pair of panel lifts with scraps of ¾-inch plywood. As you can see in the photograph (Figure 4-8), they're easy to set up.

Note the large holes in the lifts. These holes lighten the lifts and make them easy to hang up on the wall for storage back in your shop.

Door Buck

When working with doors, you need some kind of clamp to hold them upright. Commercial-made clamps are available, but I prefer to cobble some up from scraps of lumber. See Figure 4-9. Include a clamping device in your shop-made buck. A simple wedge-driven system works fine.

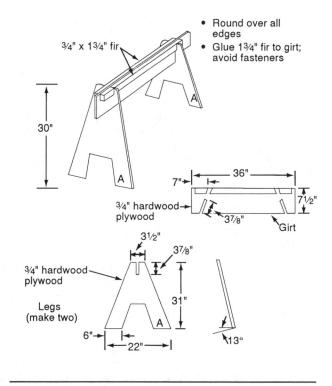

Figure 4-7   Shop-made sawhorses

Figure 4-8 **Plywood lifts set up on floor**

A lot of carpenters prefer fiberglass because it's both rugged and lightweight. And weight, as you have no doubt realized, is an important consideration when you're constantly moving a tool from place to place.

Some carpenters, though, like the "feel" of the traditional, heavier wood ladder. The additional weight gives the ladder more stability, and that's especially important when you're up there on the top two rungs.

Rent scaffolding as you need it. Always check and make sure the framework is in good shape. There shouldn't be any struts missing and the connectors should all be intact and in good working order. Make sure the planking is of scaffold-grade stock and in good condition. We'll cover scaffolding in detail in Chapter 10.

Ladders and Scaffolding

I like to have two stepladders on hand for finish work, a 2-footer and a 6-footer. Whether they're made of fiberglass or wood is up to you.

For now, though, it's time to gather up the fixtures and tools and head off to work — prep work. That's the subject of the next chapter.

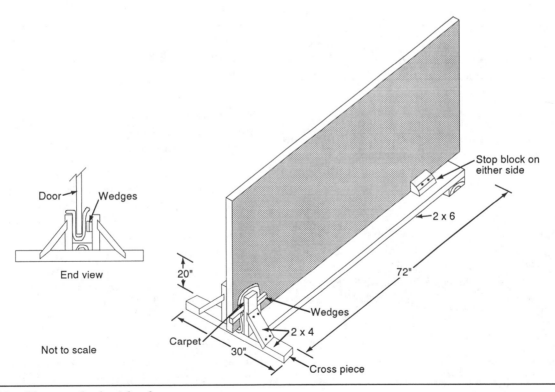

Figure 4-9 **Shop-made door buck**

Chapter 5

Prep Work

It would be great if we finish carpenters could just walk onto a job site, unpack our tools, and start hanging doors, cutting and fitting moldings, or doing any of the dozens of other things that are part of our work. Ideally, we'd have easily-accessible electrical power off the grid (no generators!). The materials we needed to work with would be carefully stored on site in a handy location. The work of other tradespeople would be out of the way, or scheduled to begin after we're through.

In this perfect world all the windows and door frames would be square, level, and free of warp. Walls would run flat and plumb and meet squarely at the corners. Blocks would have been fixed into the framing to back up the molding runs and ends. And of course the drywallers will have left the drywall off the back of the stair carriages. Dream on!

As you may have noticed by now, it's not a perfect world. There is much to be done, and often, much to be fixed, before you can get down to work. In this chapter, we'll look at the

paperwork side of prep work. The next chapter deals with the preparation of the rough framing you might want to do before the arrival of the drywallers.

Site Inspection

Can you get there from here?

Believe it or not, that's a serious question. It's easy for an owner or general contractor to say, "Access? No problem!" Then you find yourself slogging across a hundred yards of mud every morning carrying half a ton of tools — and don't forget that trip back!

Before making a commitment to one of those "No problem!" jobs, check out the site. You just might want to adjust your bid or negotiate an extra charge to cover dealing with access hassles.

When making that site inspection, check availability of power and storage facilities. Make a note of any problems involving getting power to your secondary feed box. If the temporary power pole is a good distance from the finish work, plan to get power through an adequate feed line. If power is supplied from a generator, make sure you have surge protection for your tool motors.

No professional carpenter likes installing finish materials that have been damaged in storage. I've found it's well worth my time to find a dry, out-of-the-way storage area for my materials. See Figure 5-1. Find a way to keep the

Figure 5-1   **Protected storage for finish material**

stock off the floor. On cement floors, get someone to lay a temporary vapor barrier of plastic or builder's felt. This is especially important if the cement has been recently poured.

When the materials are delivered, make sure they're covered with cardboard or builder's paper. And don't let other trades store their materials on top of your pile!

Job Sequencing

There are two phases to finish work:

1) The first stage I call the "prep work." You make corrections, additions and modifications to the rough frame. (That's explained in the next chapter).

2) The second stage is the actual cutting and installing of the finish materials.

Ideally, the second phase follows after plumbing and electrical are roughed in and the drywall is hung, taped, sanded and primed.

With luck, you'll finish the second phase before the carpet and hard flooring installers, the painters, and the wallpaper hangers arrive. You especially want to have it done before fragile lighting and plumbing fixtures are installed.

Depending on how complex the molding fits to cabinetry and other built-ins are, you may want cabinets and built-ins installed before running the moldings.

Make a list of all the things you're going to do in the second phase. Put them in order. In general, I like to work from the top down. Literally. That means ceiling and cornice work comes first, followed by standing moldings and wall treatments. And last, the base moldings. Fragile items, such as fireplace surrounds and balustrades, I usually do at the very end.

Selecting the Materials

It often falls to the finish carpenters to estimate, select, and order finish carpentry materials. Of course, the owner or the general contractor may prefer to order materials. It's their choice. But if you do the ordering, be prepared to make informed choices on wood species for both paint-grade and stain-grade application.

Knowing how to estimate the amount of running and standing molding for the project is also a part of that job. So is making up a bill of materials.

Species Considerations

The first consideration in selecting the right wood species for your finish materials is whether the surface will be painted or stained. You choose a paint-grade species for its lack of grain, ability to hold paint smoothly, and economy.

If the wood is to be left natural, the most important things to consider are grain and color.

In any case, always make sure the wood has been properly seasoned for use in residential interiors.

Attributes of Common Woods for Finish Work

Here's information on the various species, including a man-made product, commonly used for finish work.

MDF sheet stock- This is a man-made wood product composed of compressed wood fibers and glue. It's listed here because it's an increasingly common substitute for solid wood in paint-grade applications.

MDF sheet stock mills easily and smoothly, leaving crisp edges and surfaces. It glues and fastens securely and takes paint exceptionally well. It's stable and durable, but because it can absorb water, it's not a good choice for a high-moisture environment such as a bathroom baseboard.

MDF sheet stock is made in sheets and is available in lengths from 8 to 10 feet. The cost per board foot is less than other finish lumber.

Pine, western maple, hemlock, poplar- I listed these woods together because they are the most common choices for paint-grade applications. They're relatively low-priced (assuming the pine is finger-jointed — solid lengths of clear pine are expensive).

These species are fine grained, with almost no grain pattern to show through the paint. They take shapes well, and glue and fasten easily. All of them are stable, even at widths of 12 inches or more.

Red and white oak- For stain-grade applications on building interiors, these two varieties of oak, especially the red oak, are an excellent choice. They are chosen for their dramatic graining and their color. White oak has a brownish hue and red oak is definitely red.

The oaks mill well, though dull shaper blades mar the crisp edges and split out the grain. Oak is hard enough to require pilot holes for hand-driven finish nails. It's stable in narrow widths and glues well enough. It's also one of the more expensive woods.

Cherry and other fruit & nut woods- Along with pecan, butternut, and walnut, cherry is chosen primarily for its richness of color. They're all hard, like the oaks (except butternut, which is a little softer), and they take shapes just as well, if not better.

Compared to oak, these woods are very expensive. But don't let the price intimidate you. You'll find these woods no harder to work with than the oaks. In most cases you'll notice little difference.

Estimating Running and Standing Moldings

Traditionally the term *running molding* refers to horizontal finish elements, such as base, crown and chair-rail moldings.

The term *standing molding* refers to finish elements that are applied vertically, such as door jambs, cases, and window trim.

I've found that it's more convenient to use the term *running molding* to refer to any item that can be ordered by the linear foot. This type of stock comes in random lengths rather than specific lengths for a specific application.

Under the category *standing molding* on my bill of materials, I list any molding that has to be ordered in specific lengths or in packaged units. It's best to keep separate lists of running molding and standing molding when estimating the materials needed for a job.

Unless certain items have to be ordered days or weeks before delivery, I prefer to estimate the quantities needed by measuring distances on the site. Of course, you can also estimate quantities by scaling dimensions on the architect's plan.

Figure 5-2 shows the forms I use to list finish materials needed for a job. This is usually called the *bill of materials*. I tally the running moldings ordered by the linear foot. I keep a separate list of moldings and other stock items that must be ordered in specific lengths or by the square foot. Some jobs may require other categories of finish materials, but this form takes care of the basics.

Ordering Notes

Allow an extra 15 to 20 percent for waste on materials ordered by the square foot. On running moldings, a 10 to 15 percent waste allowance ought to be enough.

Be sure to note the grades of your materials clearly on the bill of materials. For example, if plywoods need to have good faces on both sides, you've got to specify it. On solid woods (other than pre-dried moldings) you've got to specify kiln-dried with a moisture content designation of MC 12 or MC 10. Higher MC numbers, or the designation "S-dry," are unacceptable for interior finish work.

Running Moldings

Standing Moldings

Type molding	Number of specific lengths or square footage										Total SF of units
	Kitchen	Dining	Living	Den	Bath 1	Bath 2	Bed 1	Bed 2	Bed 3	Misc.	
Win. case											
Stool											
Apron											
Ext. jamb											
Door case											
Blocks											
Plinths											
Valence											
Wainscot											
Panel											
Ceiling strip											
Closet pole											
Closet shelf											

Running Moldings

Type molding	Lineal footage										Total LF
	Kitchen	Dining	Living	Den	Bath 1	Bath 2	Bed 1	Bed 2	Bed 3	Misc	
Base											
Base shoe											
Chair rail											
Bed mold											
Picture rail											
Crown											
Quarter round											
Furring strips											
Corner backing											

Figure 5-2 **Bill of materials**

Non-molded solid stock also has to be grade specified. For most finish applications, a minimum grading of "Select and Better" (the National Hardwood Lumber Association rating) is generally recommended.

When ordering finish materials, order the other supplies you'll use for the job at the same time. This assures you'll get everything on the site at the same time — and even better, you'll get everything on the same bill. There's no need to make up a separate bill later.

Supplies for Finish Work

Here's a checklist of supplies to have on the job site for finish work:

1) Fasteners, including finish nails, small-headed trim screws, a selection of drywall screws, bolts for stair work, and specific fasteners for unique installations.

2) Adhesives, including carpenter's glue, panel adhesive, contact cement, and other glues for specific applications.

3) Sandpaper.

4) Fill materials to fill nail holes, pores, and minor imperfections in paint-grade materials.

5) Wood putty to fill holes and imperfections in stain-grade stocks.

6) Caulk to fill gaps in joints and between moldings and wall surfaces. Check first to see if the painters will be doing their own caulking. Even then, most carpenters like to at least fill imperfections in their joints to assure a quality job. Be sure, though, that the caulk is compatible with the paint.

7) Shim stock. Because cedar splits so easily and falls out of the gap it's supposed to fill, specify pine shim stock if available. Shim stock must be dry, or it may shrink, split and fall out of place.

Chapter 6

Preparing the Rough Frame

On nearly every job, you're going to have to make some adjustments to the frame before you can start the finish carpentry. Without these adjustments, it's nearly impossible to do first-quality work in a reasonable time. I call this part of the job the prep work. It isn't really finish carpentry. But it's an essential part of the finish carpenter's trade.

The hardest part of this prep work may be getting to the site at exactly the right time — when the framing is finished but before the drywall is up. If you can't do the prep work yourself when it should be done, you may be able to get the framers to do what's required for you. But don't count on it — they have their own deadlines and goals. And your being on site at exactly the right moment doesn't happen often either. That's why I call this the "what should have been" chapter.

Adding Blocking

You'll nearly always want some extra backing for molding running into the corners of the rooms. While the framers add a corner stud as a nailer for the drywall, it often doesn't extend out far enough to catch the ends of cornice and base moldings.

Adding extra backing at the corners is easy (as long as the drywall isn't up). Just add a block to the lower and upper corners as shown in Figure 6-1. If you're installing chair or plate rails, remember to add blocks where these will intersect at the corner as well.

Add more blocks at the base of door jacks. If you're installing wide standing casings or plinths, this blocking is essential. Otherwise the ends of the baseboards will have no backing except the wall shoe. Figure 6-1 shows door jack blocking and corner blocks.

Here are some other spots that need backing.

■ Check the framing where cabinets will be hung or where packaged molding will be applied. Add backing to support any fireplace surrounds. Wider trim may require additional blocking in some areas.

Figure 6-1   **Corner and door jack blocking in a stud wall frame**

■ Be sure there is secure backing where rosettes accept the ends of stair railings.

■ Install blocking for bath accessories such as towel bars and toilet paper holders.

Adding Running Backing

Adding an additional block here and there is usually enough. You'll seldom need running backing (2 x 4's set between the studs) unless studs are wider than 24 inches on center or unless butt joints will unavoidably fall between stud bays.

Occasionally, however, running backing is needed to do a really first-class job. For example, running backing lets you nail chair rail, picture rail, wide baseboard, or base cap snug to the wall at any point along the run. Install running backing at hip height behind wall paneling to give the panels a feeling of extra rigidity — and quality.

Strongbacks

There will be times when you have to level a ceiling where the joists are badly out of alignment. One way to do this is to attach a strongback to the joists. See Figure 6-2. Make the strongback by joining a 2 x 8 standing on edge to a 2 x 4 laying flat. Nail the 2 x 8 to the 2 x 4 securely before setting the strongback on the joists. Then drive screws through the 2 x 4 into the ceiling joists, pulling the joists flat against the 2 x 4. Of course, you can use a strongback only in unfinished attic space.

The truth is, finish carpenters don't care much about how flat a ceiling is unless we have to install something against it — and then only when that something requires critical joinery (such as frame and panels or applied beams and purlins). For those types of installations, it pays to take the time to level the ceiling.

Adding a strongback is a last resort. Generally, you won't need strongbacks if you ask the framers to crown all the ceiling joists the same

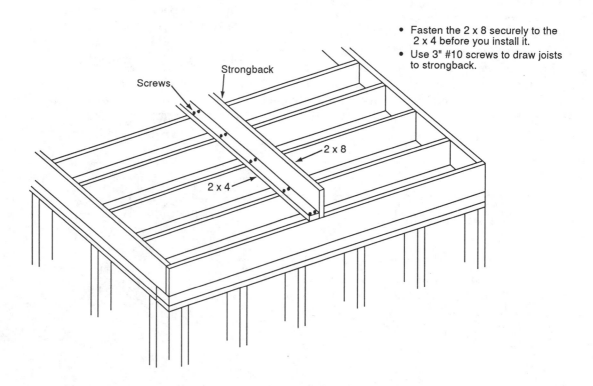

- Fasten the 2 x 8 securely to the 2 x 4 before you install it.
- Use 3" #10 screws to draw joists to strongback.

Figure 6-2   **Installing a strongback**

way — crowned upward — and to reject any "S" curved members. That should make the ceiling flat enough to meet most of your needs.

Installing Furring Strips

There's another way to bring a wall or ceiling surface flush to provide a good surface for paneling or other cover. Simply install furring strips. Furring strips provide plenty of nailing surface at convenient intervals.

For furring, use well-dried softwood that nails easily without splitting. Run the furring perpendicular to the joists or wall studs, spacing it on 16-inch centers. Insert shims between the furring strips and the backing surfaces to bring the strips into a uniformly flat plane.

There's a right way and a wrong way to insert shims. Check Figure 6-3. Note that two nails hold the furring strips in place at each joist. To keep the strips from twisting, insert shims in pairs from opposite directions. To determine how much shimming you need, stretch strings and sweep them across the plane of the furring. Or

hold a long straightedge against the strips. With a little help from your friends, you can even use a builder's level. It's slow, but absolutely accurate.

Bringing Stud Walls Into a Flat Plane

Where you'll apply wainscot, paneling, or even a simple chair rail, the wall has to be as flat as possible. If you're lucky enough to get there before the drywallers, stand at the end of a wall run and sight along the studs. Are they parallel and aligned to the shoe and plate?

Figures 6-4, 6-5, and 6-6 show three ways to straighten studs that are bowed. In Figure 6-4, I've made a saw kerf about three-quarters of the way through the stud. Driving in the shim, as shown here, bows the stud out toward you. Break off the excess shim with your hammer.

Figure 6-5 shows how to run a screw across the kerf, bowing the stud away from you. Use this as an alternative to the wedge method if, for some reason, you can't cut a kerf in the far side of the stud.

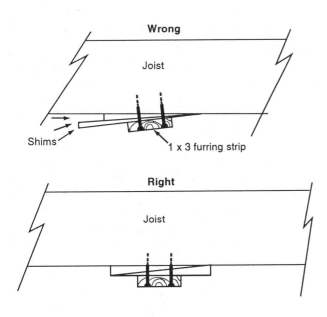

Figure 6-3   Shimming the furring strips

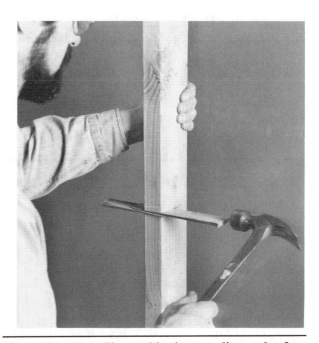

Figure 6-4   Installing a shim in a stud's saw kerf

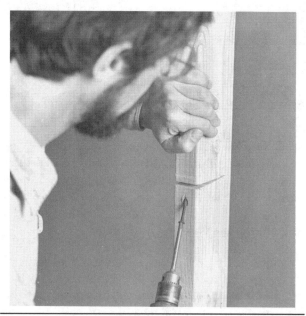

Figure 6-5　**Running a screw through a stud's saw kerf**

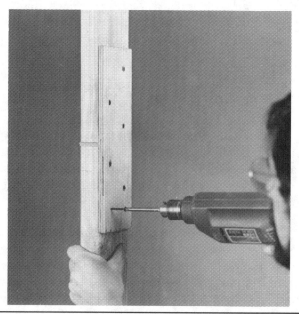

Figure 6-6　**Reinforcing a kerfed stud with a plywood scab**

Figure 6-6 shows a scab of plywood added to the side of the stud for reinforcement. You'll do this after using the shim or screw to straighten the stud.

If a stud is so badly warped that it would take more than a saw kerf to bring it flat, it's best to just remove it and replace it with a good stud.

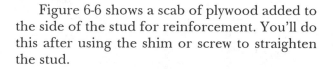

Inspecting and Correcting Door and Window Frames

Finish carpenters tend to be pretty finicky. We expect rough openings to be plumb and level and free of warp and bows, or at least close enough. If they're beyond certain limits, we have to make changes or risk a nightmare of a job installing trim moldings.

But before making any corrections, check the dimensions of the rough opening against the window and door schedule on the plans. The rough opening size for each window and door unit should be as specified by the manufacturer.

Here's a rule of thumb for most interior doors. The rough opening should be the width of the door plus 2 inches, the height of the door plus 2 inches plus the height of the finished flooring. Rough openings for wood windows should be 2 inches over the sash width. Metal windows are sized about 1 inch over the sash width.

Window height depends on the construction of the window jamb and sill arrangement. As a rule, the rough opening height must be at least $3/8$ inch greater than the total height of the window unit. This gives you plenty of room to adjust the window for level. In any case, always check and follow the manufacturer's recommendations.

While you're checking the size of the window openings, be sure that exterior building paper has been smoothly folded into

Figure 6-7   **Using a story pole to mark an opening**

Figure 6-8   **Using pinch sticks to check for a square window frame**

the openings. Make sure there are no gaps and no bulging overfolds that might interfere with setting a window.

Also check the height of the rough sill. Sill height above the floor should be uniform throughout the building for windows of the same size.

Some finish carpenters go so far as to mark the position of the window unit on the trimmers and ask the installers to level the windows to this mark. That should make the head casings the same height throughout a room. Use a story pole (Figure 6-7) with a mark at the position of the inner head jamb. Hold the pole against the floor and tight to the framing when you mark the head casing height. A story pole can make for swift and accurate window layout.

Then check the rough openings for plumb, level and square. You'll need a level, a set of pinch sticks (Figure 6-8), and a pair of string lines. Use the level to determine if the window sill and door headers are level and the trimmers are plumb.

Use the pinch sticks to determine if the diagonals are equal in the window frames. Push the sticks into one set of corners, squeeze the sticks together, and then remove them. Place the sticks in the opposite corners. If the window frame is square, you won't have to adjust the sticks to make them fit.

The frame doesn't have to be absolutely perfect. Window and door units set freely in their openings can be shimmed to plumb, level and square. But make adjustments if the framing reads more than 1/4 inch out of level or plumb over a 4-foot run, or more than 1/2 inch out of square according to the diagonals. Otherwise, setting the doors and windows could turn into another nightmare.

Pay special attention to trimmers at the door frames. I like these studs perfectly plumb, at least on the hinge side. If they aren't, I plumb the opening by driving shims between the trimmer and the king stud (which sits adjacent to the trimmer). Secure these shims to the king stud with pairs of 3-inch drywall screws spaced 6

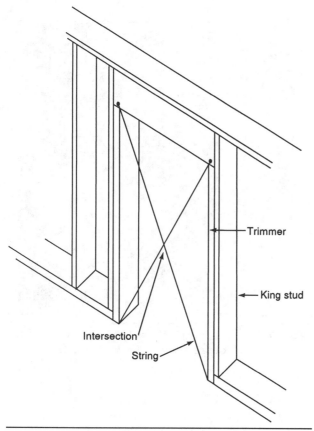

Figure 6-9   **Using diagonal strings to check an opening**

hammering one of the wall shoes forward or backward. Driving a double toenail usually brings the wall into alignment. See Figure 6-10.

Sometimes the wall shoe or headers will be out of alignment with the studs, projecting out from the wall. Plane or chisel the offending wood flush to the trimmers so the framing doesn't press the drywall out of shape — and distort the casing.

Before leaving the subject of rough openings, I should say a word about pocket door openings. Pocket doors are installed before the drywall goes up. That means what should be a finish carpenter's job (installing the pocket door) often falls to the framers. We'll talk more about that problem in the next chapter. For now, just be sure to check the size and shape of the rough opening. Is it the right size for the pocket door specified? Is it plumb? Is it level and square within the tolerances described earlier: less than ¼ inch out of level or plumb over a 4-foot run, and less than ½ inch out of square according to the diagonals?

inches apart. This makes for a trouble-free, quality door installation, especially if the doors are heavy — and most exterior doors are.

Use a pair of string lines to determine if the framing around the rough opening is warped. As shown in Figure 6-9, tack the strings so they run diagonally corner to corner in the rough opening. If the opening is true and flush, the strings will just barely touch at their intersection. If the strings gap or press into one another, the opening is warped and needs adjustment.

Door jambs should be installed parallel to the outside edge of the trimmers to make installing the case trim easier. Check which way the trimmers have to be moved to bring the opening into plumb. Then adjust the door frame by

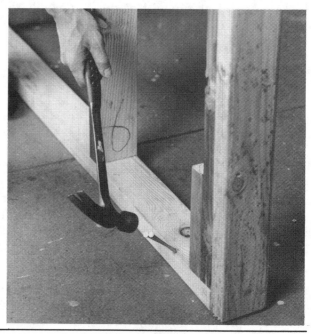

Figure 6-10   **Using a double toenail to drive a wall into alignment**

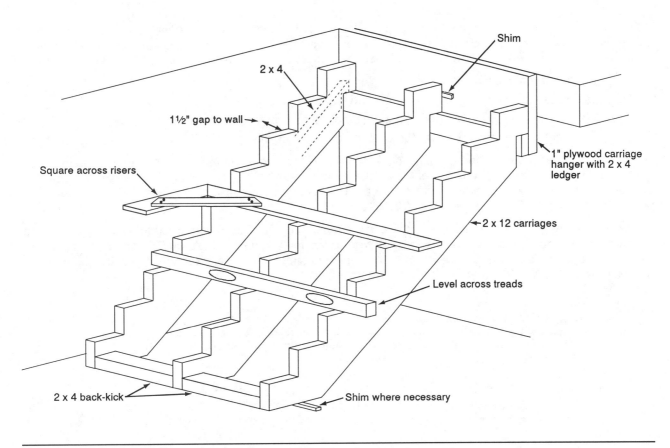

Figure 6-11 **A typical stair carriage layout with level and square in position**

Stair Carriages

I feel we do a better job of trimming out a flight of stairs if we also install the carriages. But if the carriages are already in, check to be sure they were installed correctly. Here's how:

Use a framing square to be sure the face of the risers form a right angle with the wall. Then check the level across the tread with your spirit level. Make corrections by inserting shims between the foot and head of the carriage.

Figure 6-11 shows the parts of a typical open-sided stair carriage with a square laid in position for checking the risers, and a level laid in place for checking the treads.

Before securing the carriages, be sure there's enough space between the inside carriage and the wall to allow drywall, plus a skirt board, to fit in between. Use a 2 x 4 spacer to provide a 1½-inch gap. Also be sure to install a back-kick at the base of the carriages. The framers sometimes forget it. Without it, the first riser can't be nailed securely along its bottom edge.

Finally, leave a note for the sheetrockers asking that they leave the underside of the carriage uncovered until the stair's finished treads and risers have been installed. You can nail the risers to the back of the treads and insert adjustment shims with the rock in place, but it's a lot harder to do.

Manhours for Preparing Rough Framing

Every manhour estimate depends on job conditions. No single estimate can cover all situations. But I feel the productivity figures that follow will be a good first approximation when estimating the time required for most finish carpentry tasks. All figures in the tables are in manhours and are based on the following assumptions:

- Tools and materials needed are available on site.
- The tradesman is a qualified and motivated finish carpenter.
- Work is good quality, stain grade, done no more than 9 feet above floor level.
- All defects are remedied before the carpenter leaves the site.

Add extra time for setup, cleanup, painting or staining, protecting adjacent surfaces, complicated layout or inadequate plans, repair and replacement jobs where fitting and matching is required, working around other trades, setting up scaffolding and ladders for work above 9 feet, and supervision, if necessary. Paint-grade work will usually reduce the time needed by from 20% to 33%.

Of course, every manhour estimate in this book is a poor second choice compared to figures you develop yourself on your jobs based on work done by your tradesmen. The most reliable guide when estimating your jobs will always be your own experience. But if you have to supplement that experience with the judgment of others, I expect the figures below will be worthy of your trust.

Blocking and Backing

These figures assume that blocking or backing is added before drywall is hung.
Per piece of blocking or backing set

2 x 4 backing or blocking	.04
2 x 4 strongback screwed to 10' to 16' long ceiling joist	.35

Furring

This figure assumes that furring is nailed over studs and adjusted with shims to bring the wall surface flush. Most walls will require very little shimming. More time will be required if the wall plane is very uneven or if the furred wall has to be exceptionally even. For example, frame and panel ceilings with beams require furring that varies no more than 1/32" from top to bottom and from end to end.
Per 100 linear feet of furring installed

1 x 4 furring with little or no little shimming required	1.00

Straightening Wall Studs

Cut saw kerf and drive shim or screw into the kerf. Reinforce with a plywood scab, per stud straightened	.30
Remove and replace a badly warped stud, per stud replaced	.25

Chapter 7

Installing Doors

This chapter explains how to hang the types of doors common on most residential jobs. We'll cover the standard hinged passage and entry doors, as well as sliding and bifold closet doors, pocket doors, and the French doors popular in more expensive homes. How you hang the door depends on the type. I've covered each type of door in a separate section.

The Parts of a Door Assembly

Before we get down to specifics, let's talk about the parts of a typical door assembly. Figure 7-1 shows the parts of a frame and panel interior passage door. Notice that the jamb head is dadoed into the side jambs. Sometimes this joint is a simple rabbet.

Notice also that the jamb edge is back-beveled away from the wall surface. This assures that the casing won't be held away from the edge if the jamb isn't at a perfect 90-degree angle to the wall.

If you're going to hang doors for a living, take the time to understand what's meant by the hand of a door. Obviously, you have to know where to install the hinges so the door swings the right way. The hand of a door tells you that.

Luckily, there's an easy way to figure out the hand of a door. Simply imagine yourself standing in the door frame with your butt against the hinged (or butt) side of the frame or jamb. If the door swings away from your right hand, it's

right-hand door; if it swings away from your left hand then — you guessed it — it's a left-hand door. I call this the butt-to-butt rule.

Now that you know the rule, I've got some bad news. Some door hardware manufacturers don't follow it. To some manufacturers the hand of a door refers to the side on which you place the locking mechanism. So, while most hardware companies (Baldwin, for example) follow the butt-to-butt rule, don't count on it. When in doubt, check with your hardware supplier.

Before starting to hang any door, check the frame. Is the rough opening correct? Is it reasonably plumb, level, and free of warp? If you need to correct any of these things, Chapter 6 tells you how. Next check the swing direction on the plans. Will that swing work for this opening? Don't just blindly follow the plans. Architects make mistakes just like everyone else.

The Four Rules of Door Swing

1) The swing should be *away* from traffic areas, such as hallways, and into enclosed rooms.

2) Closet doors should open *into* rooms.

3) The swing should be *away* from the light switch side.

4) Doors at a room corner should swing *against* the wall, not into open space.

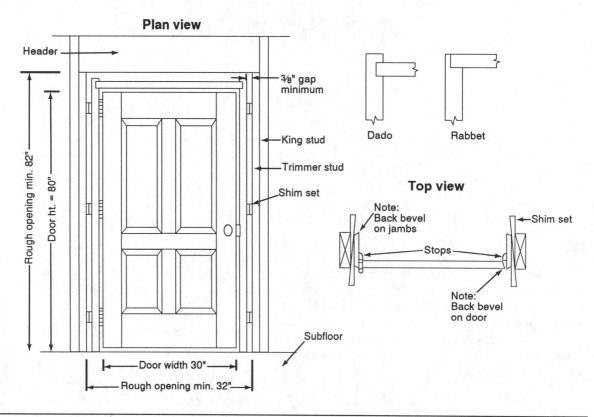

Figure 7-1   **Passage door views**

Installing Interior Passage Doors

Some finish carpenters insist on hanging the jambs before cutting any door to size. However, as a practical matter, most doors arrive on site prehung in the jamb by the manufacturer. I've done it both ways. If a construction situation demands that jambs be installed before the doors are on site, I'll do it. But I'd rather wait. Installing doors and jambs separately takes more time. I avoid it when I can.

Generally, door installation begins with the assembly of the jamb, unless the doors are delivered prehung. If the casings are preinstalled, I suggest removing them. Here's why:

- You can do a better job than the manufacturer in fitting the molding.

- The preinstalled casing makes it harder to shim and secure the door to the trimmers.

I also recommend that you avoid manufactured prehung interior doors with rabbeted stops. They make it harder to correct warped doors and leave the door with head casings that appear to be at different heights on either side of the door. That's an obvious defect in a room with both right- and left-handed doors.

Preparing the Jambs

Begin assembly of the jambs by cutting stock to the proper width. That's usually the total thickness of the wall plus another ⅛ inch. Then back-bevel the edges several degrees as shown in

Figure 7-1. Stock sold as jamb material has grooves cut into the back surface to relieve warping stress. If you make your own jambs on site, three or four saw kerfs cut ½ inch apart and ½ inch deep will be enough to relieve stress in standard ¾-inch stock.

The Dado Joint

Cut the side jambs to the height of the rough opening, subtracting at least ⅜ inch at the top for adjustment. Then cut the header jamb to

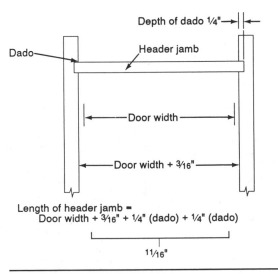

Figure 7-2   **Detail of header layout**

length. When fully buried in the dado of the side jambs, the opening should be equal to the width of the door, plus ³⁄₁₆ inch. See Figure 7-2. Cut the dado ¼ inch into the ¾-inch stock.

Use a simple shop-made dado jig to make the dado quickly and accurately. Figure 7-3 shows how you can make a dado jig from ¾-inch hardwood. The edges must be straight and the cross piece glued and screwed perpendicular to the side piece. Fit the router with the straight-sided bit to be used for dadoing. Hold the router base snug to the jig's cross piece and make the cut. The resulting dado is your reference index.

Place the dado so the inside legs of the jambs are long enough to accommodate:

- The height of the door, usually a standard 80 inches.
- The clearance from the top of the door to the header jamb, ³⁄₃₂ inch.
- The height of the flooring or carpeting.
- The threshold if specified, plus clearance to the flooring. Usually ½ inch above the finished floor is enough clearance.

Cutting the Mortises

Before assembling the jamb unit, cut mortises for the hinge plates on the hinge-side jamb. Use a plywood jig and a router fitted with a mortise cutting bit. See Figure 7-4. Notice that

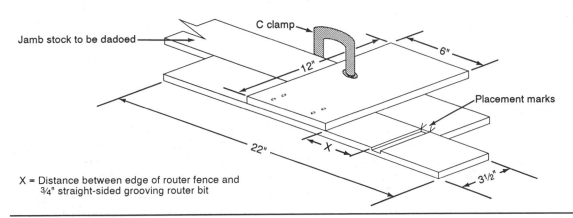

Figure 7-3   **Dado jig**

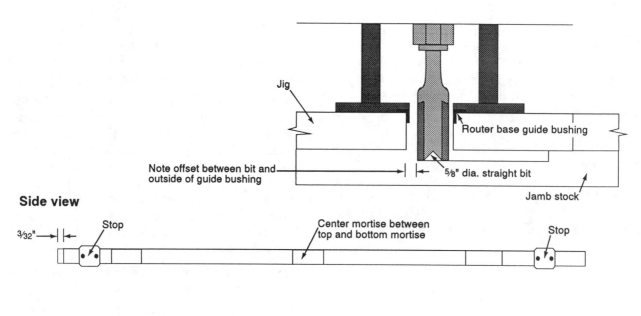

Side view

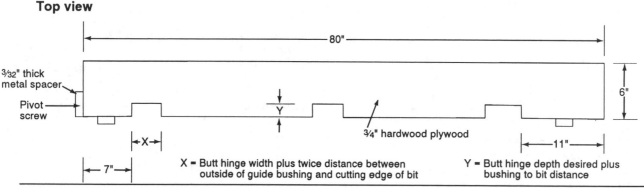

Top view

Figures 7-4   **Butt mortise jig**

there's a ³/₃₂-inch-thick spacer at the top of the jig. This allows the clearance needed between the top of the door and the jamb header. Hold the top of the spacer flush to the bottom of the header dado when clamping the jig to the jamb. Be sure the stops along the side of the jig are tight to the jamb's edge.

Cut the mortise depth equal to the thickness of the hinge leaves — except for the top mortise. It should be about ¹/₃₂ inch deeper than the others, burying the hinge slightly below the surface of the jamb. That counteracts the tendency of a door to sag away from the top hinge. This step is particularly important when working with heavy doors. I'll be mentioning it again in the section coming up on hanging the door.

Outline the shape of the striker plate on the opposite jamb with a sharp knife. Then use a hand chisel to cut the mortise. Its position should match the precut latch hole on the door. If there isn't one, the standard height for door knobs is 36 inches. Buildings intended for use by the handicapped, however, may have lower hardware heights.

You can also cut out the butt mortises by hand with a chisel and hammer. Most building supply yards carry markers for defining the cut lines of specific size butt hinges. If you can't find one, run a knife around the edge of the hinge to mark the outline. Drive the chisel across the grain and down to the desired depth about every ¼ inch over the entire area of the leaf. Then slice

Figure 7-5   **Chisel cutting butt mortise**

the chisel in from the side to remove the perforated wood. Look at Figure 7-5. It's a close-up of the chisel cutting the butt mortise.

Installing the Hinges

When the mortises are cut, separate the butt hinges and install the leaves to the jamb. Unless the hinges have rounded corners, you'll have to chisel the corners of router-cut mortises square. A corner chisel makes quick work of this task. Use a special pilot hole bit called a *Vix bit* to locate the center of the holes. See Figure 7-6. Cut a hole for the latch bolt. Then hand mortise and install the striker plate.

The spacing, size, and number of butt hinges needed to hang a door depend on the width and thickness of the door. Figure 7-7 is a table you can use to size the butt hinges.

Here's how to space hinges on a door that requires three standard pinned butt hinges. Place the first hinge 7 inches from the top of the door. Place the second hinge 11 inches from the bottom. Place the third half-way between the other two. On larger or very heavy doors, I add a fourth hinge to support the weight or use 5-inch to 6-inch ball-bearing pinned hinges. Lightweight hollow core doors generally have only two hinges.

Installing the Jamb Assembly

Assemble the jamb on a flat, level work surface of comfortable height. A pair of leveled sawhorses works perfectly. Attach the side jambs to the header jamb with screws and glue, even though end-grain gluing is marginal at best. Fix a temporary stop spaced back the thickness of the door, near the striker plate. Tack a spreader bar near the bottom of the jambs to hold the jambs parallel.

Carefully lift the assembly and carry it to the framed opening. Center it between the trimmers. Double-check to be sure the hinges are on the correct side and are facing the right direction. It's easy (and embarrassing) to make a mistake: Check twice, hang once, not the opposite!

Plumb the hinge side jamb leg to both the edge and face of the door plane. The jamb should be flush to the surface of the walls. A 6'6" jamb level (like the one shown in Figure 7-9) is ideal for this job. If you don't have one, make do with a smaller level. Temporarily tape it to a straightedge cut to the length of the inside jamb.

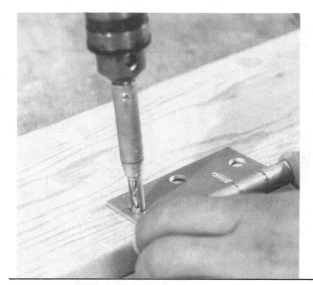

Figure 7-6   **Vix drill bit locating center of pilot holes**

Thickness of door (inches)	Width of door (inches)	Size of hinge (height of leaf) (inches)
1⅛ to 1⅜	up to 32	3½
1⅛ to 1⅜	32 to 37	4
1⅜ to 1⅞	up to 32	4½
1⅜ to 1⅞	32 to 37	5

Figure 7-7   **Sizing butt hinges**

If the wall is out of plumb, the edge of the jamb may have to protrude from the wall to be plumb. *Be sure this protrusion is at the bottom of the jamb.* At the top of the jamb it would cause a nightmare in the joining of the casings.

The jamb should also run straight and square with the wall. If it doesn't, adjust the shims until it looks right. There should be a set of shims behind each butt hinge. If necessary, shim up the hinge-side jamb slightly above the floor. That way the header can move to either side of square in relation to the hinge jamb. This helps the jamb header conform to the shape of the door once it's hung in the opening. See Figure 7-8.

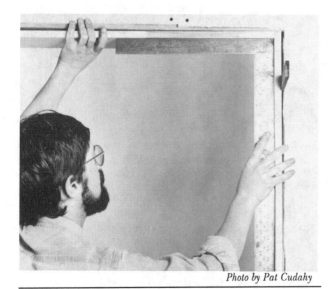

Photo by Pat Cudahy

Figure 7-8   **Square held in door jamb**

Now, holding the jamb tight against the shims, with the pressure of the jamb level, firmly secure the hinge-side jamb to the trimmer. Use 2½-inch air-driven (or 16d casing) nails set through the shims. See Figure 7-9.

Before hanging the door on the hinge jamb, check its height. The usual bottom clearance is ½ inch above the finished floor. You have to know the height of the finish floor and the threshold, if any. Be sure clearance is adequate under the entire swing of the door before figuring the height. If in doubt, leave the door long. You can always cut it down later.

If you have to shorten the door, use a circular saw to make the cut square to the edge of the door. Guide the saw with a jig like the one shown back in Chapter 3, Figure 3-2. Set the jig to a cut line scribed into the door with a utility knife. The cut line reduces wood splintering on the face of the door.

Note that it's essential to customize the jig for a particular saw and saw blade. Changing either may throw the saw blade out of the intended cut line.

Cutting Mortises for the Butt Hinges

With the door held on edge in a door buck (shown in Figure 4-9, Chapter 4), cut mortises to receive the butt hinge leaves. Tack to the door edge the same jig used to cut the jamb hinge mortises. Swing the spacer down and index it

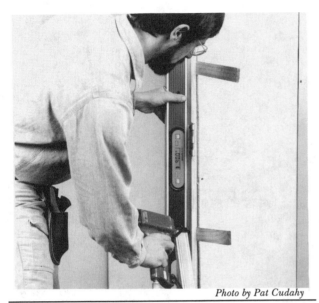

Figure 7-9   **Jamb level holding jamb against shims for nailing**

Figure 7-10   **Door buck holds door as jig guides router to cut hinge mortises**

against the top of the door. Set the jig's stops firmly against the side of the door, as in Figure 7-10. Double-check the side where the leaves should protrude. Then cut the mortises. Once the mortises are cut (and square-cornered, if necessary), install the leaves.

The door may not have a back-bevel worked into the striker side edge. (See Figure 7-1). In that case you'll have to create one. Here's how:

1)   Clamp the door striker-side up in the door buck.

2)   Draw a mark on the side of the door to which the bevel will slope. The mark should indicate the cut depth, and depends on the degree of the back bevel: 2 degrees for 1⅜ inch doors and 3 degrees for 1⅝ inch doors.

3)   Carefully remove the wood down to this mark using a power planer or a hand plane.

Installing the Door

Remove the temporary spacer bar from the bottom of the jamb and bring the door to the jamb. Starting with the top or bottom hinge (I want no part of this argument!), mesh the leaves together. Press the pins into place. If you used a router jig to cut the butt mortises, it should be a perfect fit. If it isn't, give the top or bottom of the leaf a light tap. The fingers should pop right into place.

If it's a heavy door — and most are with kids swinging on them! — I like to replace the standard screws on the hinge leaves holding the door to the jamb with longer ones. Use screws long enough to bury at least ¾ inch into the trimmer.

With the door in place, adjust the jamb to run even with the edge of the closed door. Raise or lower the striker-side jamb so there's an even margin between the top of the door and the header jamb. I use wedges to raise the jamb. (See Figure 7-11). When you're satisfied with the

Photo by Pat Cudahy

Figure 7-11   **Hammer wedges under jamb bottom to adjust alignment of top of door with header**

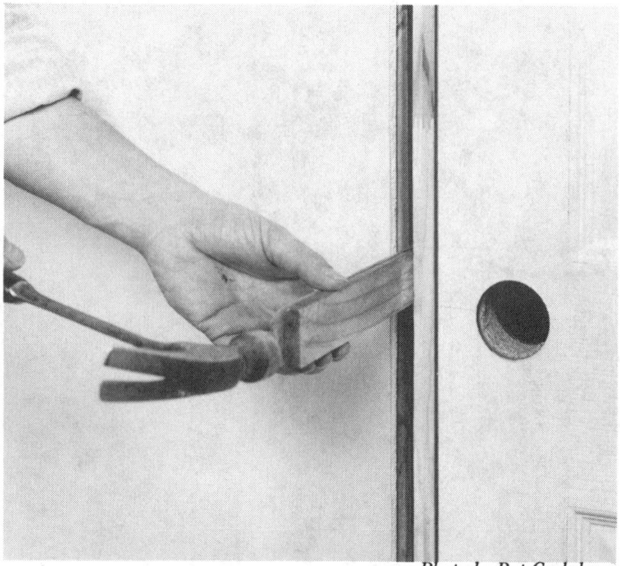

Photo by Pat Cudahy

Figure 7-12   **Adjusting margin line along striker side of door by setting shims**

margin and you've got the upper portion of the jamb flush with the wall surfaces, attach it at the upper corner. Nail it through a set of shims to the trimmer.

To adjust the margin along the length of the strike-side jamb, move the shims wedged between the jamb and the trimmer. When adjusting these shims, remember to keep the jamb plumb and square to the plane of the wall. What you want here is an even gap the width of a dime running the length of the door. (See Figure 7-12). Checking and final-setting the striker jamb will require constant opening and closing of the door to adjust the shims.

If I'm hanging heavier doors, I like to sink the top butt mortise a little deeper into the jamb, as I mentioned earlier. This gives a taper to the margin along the hinge edge. Reproduce this taper along the striker edge. Then, when the door sags away from the upper hinge corner, you'll get a straight line all around. The nice thing about this theory is that it generally works.

Installing the Latch Mechanisms

Unless the door was precut for a latch mechanism, you'll need to cut the holes yourself. Do it after you've hung the door. It's easier and just about foolproof. (Why open yourself to the possibility of making a mistake if you don't have to?)

1) Find your height position by marking the centerline of the installed striker plate along the face of the door. With a combination square, transfer this line to the edge and to the opposite face.

2) Open the door and secure it by driving a set of shims under the bottom.

3) Determine the size of holes and setbacks with the template supplied by the latch manufacturer.

Nothing to it so far, right? A specialty jig made to guide the auger bit into the door makes the rest just as easy. You can also use a shop-made

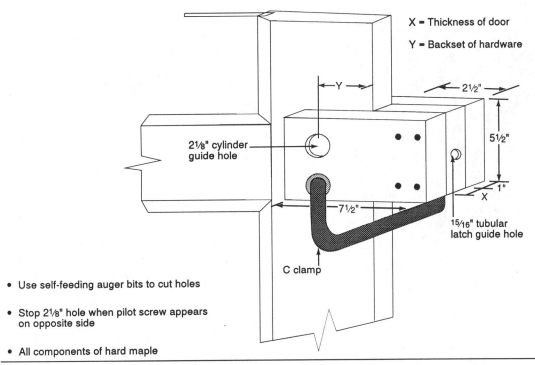

X = Thickness of door

Y = Backset of hardware

2⅛" cylinder guide hole

2½"

5½"

1"

X

7½"

¹⁵⁄₁₆" tubular latch guide hole

C clamp

- Use self-feeding auger bits to cut holes

- Stop 2⅛" hole when pilot screw appears on opposite side

- All components of hard maple

Figure 7-13   **Shop-made hardwood jig for installing latch mechanisms**

hardwood jig. (See Figure 7-13). But you'll find you'll eventually need to replace it as the holes enlarge with use.

4)   Drill the large hole for the knob levers through the face of the door. Don't cut these large holes all the way through. You're apt to get nasty tearout if you do. Stop when the drill's pilot screw just breaks through. Now finish it off from the opposite side, centering the bit on the pilot hole.

5)   Next drill the smaller hole for the tubular latch mechanism.

6)   Now mark the mortise for the latch bolt plate by sliding the tube into the hole. Outline the plate on the door edge with a knife. Remove the wood by hand chiseling or make up a jig like the butt mortise jig, and use a router.

Installing the Door Stops

Here's how you install the stops around the inside perimeter of the jamb:

1)   Install the header stop first. Use pinch sticks (as in Figure 7-14) to determine the correct length. If the stop has a molded edge, miter the ends to length. Otherwise, crosscut the ends square. When installing the stop, hold it tight to the closed door on the striker side. Hold it the thickness of a dime away from the door on the hinge side.

2)   Cut the hinge side stop to length. Install it a dime's width away from the closed door along its entire length. This keeps the back of the door from binding against the stop as it swings open.

3)   The striker side stop is a little different. While the top and bottom of the stop are tight to the closed door, the area between is held slightly away. I go up to

Photo by Pat Cudahy

Figure 7-14   Use pinch sticks to determine length of header stop

about the thickness of a nickel near the striker plate. This trick introduces a bit of spring to the opening action of the door. It also exerts pressure against the latch when the door is closed. This prevents rattling.

With the striker-side stop, you'll have to experiment some to determine the amount of gap for the type of door and latch mechanism you're working with. Too much spring will prevent the latch from engaging the striker under normal closing pressure. Back off pressure by lightly tapping the top and bottom of the stop with a block and hammer. Move the stop slightly away from the door.

Installing Pocket, Bifold, Sliding and French Doors

Today's finish carpenters have to know more than just how to hang the standard hinge-swung passage door. I mentioned back at the beginning of the chapter that we'd cover nonstandard doors in a separate section. This is it.

The bad news is that installation can vary quite a lot from type to type. For example, extra stops and hardware items are often necessary. Valances may be needed to hide the hardware. Some bifolds and sliders have no jambs at all. Other doors can't be prehung before the jambs are installed. The good news is that it's not as bad as it sounds.

Pocket Doors

I've found the biggest trick to hanging pocket doors is to get there before the framer does. Pocket doors have to be installed before the drywall goes up. This means the framers often end up installing the jambs. And if they're not put in right, they may never work properly. So be on the site first. Make sure they get *installed* — not slapped in place!

The hardware setup lets you adjust the fit of the door to the jamb when the door is in the closed position. The problem is that it can't simultaneously adjust for the fit in the open position (Figure 7-15). For both lines to be true, the header jamb must be installed straight and square to the side jamb.

Sequence for installing a pocket door:

1) Assemble the frame. Cut off the protruding ears on the striker (open) side.

2) Center the assembly in the opening. Attach the header jamb to the frame header. Use shim sets and a level to make sure you've got a level and flat installation.

3) Use your level to plumb and straighten the side jamb. Attach it to the trimmers through the shim sets. Use a framing square to double-check the corners for square.

4) Attach the metal flange or wood strip in the pocket area of the door frame to the floor with screws.

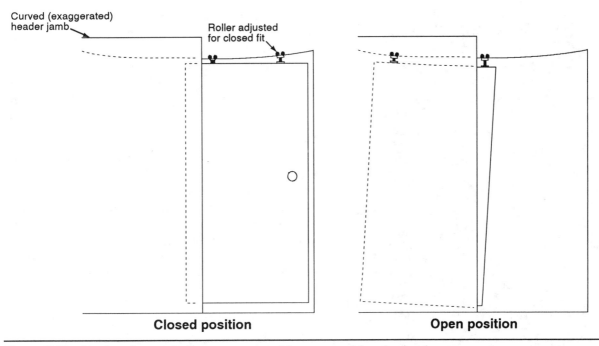

Curved (exaggerated) header jamb

Roller adjusted for closed fit

Closed position   **Open position**

Figure 7-15 **Pocket door misalignment caused by a curved header jamb**

5) Install the pulls and locking hardware.

6) Hang the door and install the valances that hide the header hardware. Adjust the roller hardware so the door closes evenly to the striker jamb. Finally, hold the door plumb with wedges between the door and the floor and install the vertical stops.

Bifold and Bypass Doors

On bifold and bypass doors, make sure you size the jambs according to the recommendations of the door hardware manufacturer. Inside width and height depend on the door size plus the recommended clearance for the hardware. Clearance will vary among manufacturers. *Make sure you check!*

Jambs must be installed before the doors are hung. Be sure you get the side jambs plumb and perfectly parallel to one another. I do it this way:

1) Carefully plumb one side with a jamb level. Choose the side that's sitting on the high end of the floor. That way you won't have to cut down the other side jamb to get the header level.

2) Use a set of pinch sticks to measure over and set the other jamb parallel to the fixed jamb.

3) Make sure the header jamb is installed flat and level so the doors will close gap-free to either side jamb.

Here are a couple more things to look out for on bifold and bypass doors:

- The hardware at the bottom of the door (pivots for bifolds and bypass separators for bypass) must be blocked up from the subfloor.

- If a hardwood floor is to be laid, leave the hardware off until it's in.

- For carpeted floors, install the hardware over plywood blocks that simulate the thickness of the carpet and pad. Shape the blocks to the outline of the hardware's fastening plate.

Figure 7-16 **Modern entry door installation**

French Doors

Installing French doors is basically like hanging two passage doors at the same time — though in other ways it's like hanging bypass or bifold doors.

Like bifold doors, make sure the header is perfectly flat and level. The side jambs must be plumb, parallel, and spaced to accommodate the two doors plus clearances (door widths plus $9/32$ inch).

Before nailing off the jamb of a French door, hang the doors on their hinges and check for proper closure. You want an even gap about the thickness of a nickel between them.

If jambs are installed correctly and the doors are square, there should be no taper to the gap and no need to readjust the jambs. Correct a slight taper by planing the edge of one door. You can also make taper corrections by shimming hinge plates out from the side jambs. Insert thin strips of card stock between the hinge leaf and the mortise.

If there's a warp in one or both doors, or an out-of-plumb side jamb, one door will protrude in the closed position. It's almost impossible to correct a warp in the door itself. So try moving the bottom of a side jamb to match the warp in the door. If you need a lot of movement, move both jambs in opposite directions. If the warp is bad enough, you'll have to replace the door.

Installing Exterior Doors

In one way, installing exterior entry doors is like installing pocket doors — the finish carpenter had better get there before the framer does! Especially if the door includes side lights, which makes proper installation even more critical.

If you ever have to trim out one of these beasts, you'll understand why it's so important to get the door and framing level, plumb and flush. A sloppy framer's work may have to be repaired before work can begin.

For the most part, installing exterior doors is much the same as installing interior passage doors, with these exceptions:

- Be sure you wrap the rough opening in felt or builder's paper to protect the frame from moisture. Smooth out any bulges that will interfere with the door installation.

- Exterior doors with integral wood sills may need alteration of the subfloor or even the floor framing itself so the sills will protrude the right amount above the finish floor. The extent of alterations depends on the configuration of the sill. The finish floor usually comes just to the bottom edge of the inner bevel. Luckily, installations like this are dinosaurs. Modern entries feature adjustable metal sills designed to sit flat on the subfloor. See Figure 7-16.

■ Factory-made exterior jamb stock is generally milled from thicker wood (5/4-inch stock) than interior jamb stock. Be careful when setting the side jambs to the door. There's a rabbet worked into the stock as the door stop. If the door is warped, you may have to move the striker-side jamb slightly out of plumb so the rabbet won't interfere with door closing. If you need a lot of movement, plane and chisel back the rabbet instead of bringing the jamb further out of plumb.

■ If side lights are included, install them on the door jamb first. Then hang the assembly as a unit.

■ Most entry doors have a mortise lock. You may also have to install a deadbolt. Install both according to the manufacturer's directions.

■ Unless weatherstripping is part of the jamb stock, you'll need to add it after fitting the door. The weatherstrip, rather than the door-stop rabbet, creates the "spring" action of the door.

Manhours for Door Installation

All figures in the tables are in manhours and are based on the following assumptions:

- Tools and materials needed are available on site.
- The tradesman is a qualified and motivated finish carpenter.
- Work is good quality, stain grade, done no more than 9 feet above floor level.
- All defects are remedied before the carpenter leaves the site.

Add extra time for setup, cleanup, painting or staining, protecting adjacent surfaces, complicated layout or inadequate plans, repair and replacement jobs where fitting and matching is required, working around other trades, setting up scaffolding and ladders for work above 9 feet, and supervision, if necessary. Paint-grade work will usually reduce the time needed by from 20% to 33%.

Cased Openings

Install jambs only, per opening.

Opening to 4' wide	.50
Openings over 4' wide	.75

* Installation times for casing are at the end of Chapter 8.

Exterior Doors

Based on hanging doors in prepared openings. No lockset, casing or stops included. Add extra time if the framing has to be adjusted before the door can be set.

Prehung flush or panel 1" thick solid core single doors, including felt insulation, based on an adjustable metal sill, per door

Up to 3' wide	.75
Over 3' wide	1.25
Add for door with sidelights	.75
Sliding glass doors, opening to 8' wide, per opening	1.75
French doors, 1" thick, to 6' wide, per door	1.75

* Add lockset and stop manhours from the next page. Find casing manhours at the end of Chapter 8

Prehanging Doors Off Site

Make and assemble frame, per frame	.75

Set hinge on door and frame, panel or flush door, per hinge

Using jigs and router	.16
Using hand tools	.50
Add for French doors, per door	.25

Interior Doors

Based on hanging 1⅜" hollow core doors in prepared openings. No locksets or casing included.

Prehung interior doors, per door 50

    Add for assembling a factory-cut, mortised and rabbeted jamb 20

Pocket doors, per door

    Assemble and install the door frame 50

    Install the door and valances in an existing frame 50

Bifold doors installed in cased openings. Cased opening manhours are on previous page.

    Single bifold door to 4' wide, per door 50

    Double bifold door, to 8' wide, per pair of doors 75

Bypass closet doors installed on a track in a cased opening. Cased opening manhours are on previous page.

    Double doors, per opening 66

    Triple doors, per opening 1.00

Rehanging most hollow core doors in an existing opening, per door . . 1.50

Rehanging large or heavy solid core doors in an
existing opening, per door 1.75

* Add lockset manhours below. Find casing manhours at the end of Chapter 8.

Lockset Installation

Set in either a solid core or hollow core door, including setting the strike plate, per lockset

    Standard 2⅛" bore, factory bored .16

    Standard 2⅛" bore, site bored 1.00

    Mortise type lockset, using a boring machine 1.50

    Mortise type lockset, using hand tools 3.00

Pocket door hardware

    Privacy latch 66

    Passage latch 30

    Edge pull 30

Setting Door Stop Molding

Set stop at the head and on two sides, per opening

Panel or flush door 25

French door 33

Pocket door 33

Chapter 8

Door Trim

Once the doors are installed, your next job is to cover the gap between the jambs and the wall. The molding for this is called *casing*. While there are a lot of different types of molding profiles to choose from, most modern styles are 2¼ inches wide.

In some Early American designs, the plans will specify casings cut to widths of 3 inches or more. They may have no shaped profile at all. Sometimes a separate back molding is used to give shape to the outside edge.

Besides the molding profile, a finish carpenter is given specifications about the style of the frame itself. In the U.S. we use these three basic styles:

1) The mitered frame

2) The butted head frame

3) The traditional corner block and plinth frame

Figure 8-1 shows all three styles. If one or more horizontal moldings are added to the head casing, the butted head is then called a *cabinet head*. The butted head will sometimes have plinth blocks at the base. Note the fluting on the side cases of the block and plinth frames. Like the cabinet head, this treatment is generally found in older homes and commercial buildings.

After some initial comments on trim work preparation, I'll spend the rest of this chapter describing how to cut and fit moldings for each of these frame types.

Preparation for Trimwork

Before you start trimming out the door jambs, consider if it's the right time. It's best done before installation of some moldings, but after the installation of others.

I like to install ceiling treatments and cornice trim before tackling the standing moldings or any other wall treatments. Stepladders and scaffolding used during installation of ceiling trim can damage or mar standing moldings. This is another example of the "top down" theory I mentioned in an earlier chapter: ceiling treatments and cornice trim go on first, then the standing moldings. After that, install the running molding (chair rail, picture rail, and other moldings that abut the standing molding).

There are other factors to consider as well. Logically, before you begin, you want the drywall up, taped, mudded, and sanded out. Check to see that the drywall around the doors is securely fastened. If it protrudes past the jambs, use a *Surform*® plane to work it back to flush. See Figure 8-2.

Whether the casings are finish painted before or after you install them is usually up to the painter. Most new homes are spray painted. So unless the trim is to be stained, painters usually like the trim installed before they spray. A hand painter will probably prefer to have the trim installed after the walls are painted.

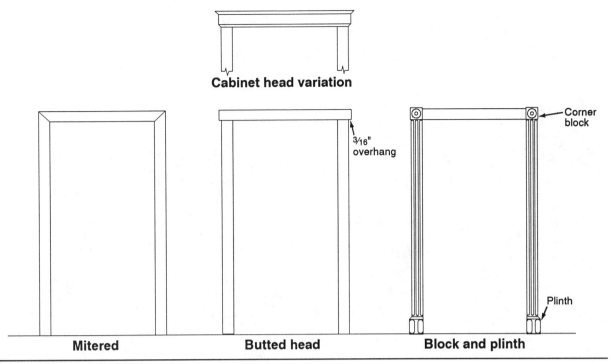

Cabinet head variation

³⁄₁₆" overhang

Corner block

Plinth

Mitered   **Butted head**   **Block and plinth**

Figure 8-1   **Door case styles**

Photo by Pat Cudahy

Figure 8-2   **Using a *Surform*® tool to smooth drywall**

Some painters preprime, or even finish paint or stain out the trim, before they hand it over for installation. Unless you plan on doing the painting yourself, find out what the painters want before you begin.

Here are two other points to consider before you begin trimming out the doors:

1) If there will be floor-to-ceiling wall paneling surrounding the doors, it's generally installed before the trim.

2) If any kind of wall paneling will be installed, check to see if you need to install extension jambs to bring the jambs out flush to the added wall thickness.

Cutting and Fitting a Mitered Frame

Begin fitting the casing to a jamb by marking the amount of reveal that will be exposed between the inside edge of the molding and the face of the jamb. I like to have ³⁄₁₆ inch of jamb exposed for a 2¼-inch casing.

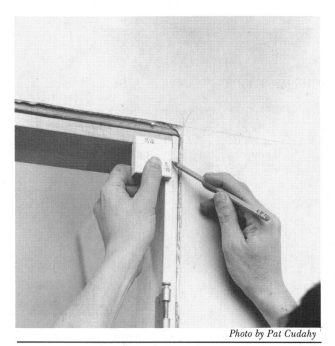

Photo by Pat Cudahy

Figure 8-3   **Use a block of wood to mark reveal line intersection**

Photo by Pat Cudahy

Figure 8-4   **Transferring the reveal intersection to the head casing**

Figure 8-3 shows me measuring and marking the 3/16-inch reveal with a block that's been cut on all edges with a 3/16-inch-deep rabbet. You could also measure and mark with a combination square set for 3/16 inch, but it's awkward, especially at the jamb corners. I like the block—it's fast and foolproof. (And I'll take any help I can get!) When I'm installing casing, I keep this block in a pouch on the left side of my work belt.

After drawing in the reveal lines, measure how long the casing has to be. I recommend that you crosscut all the standing moldings to a rough length first. Then stack each set of moldings by the door where it will be installed. In stain-grade work, grouping the moldings into sets that are grain- and color-matched is essential for a quality installation.

The rough length of each molding should be the distance between the reveal lines plus an appropriate amount:

■ For the head casing, it's the distance between the reveal lines plus twice the width of the case plus 1 inch for waste.

■ For side casings, it's the distance between the reveal line (on the header jamb) and the subfloor (or the finish floor, if installed) plus a single width of casing plus 1/2 inch for waste.

When it's time to install the casing, finish cut the head casing first. Begin by cutting the left miter as near as possible to the end of the stock. Then set the case in position, with the inside of the miter just touching the intersection of the reveal lines at the left side of the opening. Holding the case securely in position, mark the intersection of the right reveal intersection mark on the piece with a sharp pencil. See Figure 8-4. Now cut the miter to this mark.

Double-check the piece for length, then nail it to the jamb and the wall. I use a pneumatic nail gun to shoot in pairs of 1½-inch slight-headed finish nails about every 16 inches. If you're hand nailing, drive 4d nails through the inner edge of the casing into the jamb. Then drive 6d or 8d nails along the outer edge into the trimmer. (Hardwood moldings may require pilot holes for the nails.)

Working with Hand-Driven Nails

In finish work, hand-driving nails has its drawbacks. Ungalvanized finishing nails are oily and can leave a stain. Hand driving tends to split out the wood and requires two hands. But even if you have a pneumatic nail gun, there will be times when you need to hand-drive nails. Here are some tips on making the job easier:

- Dull nail points with a hammer so you get a chisel-like effect when driving. This reduces splitting.

- Use galvanized rather than common nails. Even though galvanized nails are a little thicker and leave a more unsightly hole, they reduce splitting. Even more important, they're free of the machine oil that coats most finish nails.

- When hand-driving, you'll often miss a strike or two. That leaves "grannies" in the wood beside the nail hole. These grannies aren't much of a problem in paint-grade work. Just fill the dents with a nonshrinking or cracking filler. I recommend Durham Company's *Rock Hard* water-based filler or a catalized system like *Bondo* called *Poxywood*, which is specially formulated for use on wood.

- Fixing a granny in stain-grade work is a little harder. The dent has to be raised, not filled. The easiest way to do it is with a household iron and a damp cotton rag. (See the chapter on troubleshooting for details.)

With the head casing installed, cut the miter for the top end of each side casing. Be sure to cut these sides a little long. Hold the top of the side casings against the miters of the head casing to check the fit of the angle. Since the piece is still too long for the opening, the bottom third won't be flush against the wall. See Figure 8-5.

Be sure the casing runs flush to the reveal lines. The miter joints should be perfectly matched. If not, adjust the angle of the side case miter with a few passes with a block plane.

If the jambs protrude from the wall near the upper corners, the casing won't fit flush against both the wall and the jamb. The casing will be angled back toward the wall. This opens the joint

Photo by Pat Cudahy

Figure 8-5   **Checking miter fit at the top of a side casing with bottom third of casing away from the wall.**

at the head even if the miters are perfect. Correct this problem by planing the miter angle with an undercutting bevel.

When the joint tests perfect, cut the side case to length. If the finish floor isn't installed yet, measure the distance from the subfloor to the top of the head casing. Cut the side cases so they just touch the subfloor. They'll be cut in place to the correct length by the flooring installers.

If the finished floor is already in place, cut the casing to the exact length. Instead of measuring the length, do this:

- Turn the mitered end down. Be careful not to dull the point of your miter!

Photo by Pat Cudahy

Figure 8-6   Corner block at intersection of reveal lines

- Mark the side piece where it touches the point of the header's miter.
- Draw a line horizontally across the casing and cut along that line.

My suggestion is that you cut the casings about 1/32 inch oversized so there's a little material left if some fine tuning is needed. Anyhow, a casing that's a tiny bit too long will snap neatly into place. A casing that's a little too short leaves a gap at the floor.

When you're satisfied with the fit, nail off the side cases the same way you did the head casing. Nail across the miter joint from the outside edges of the molding. I like to apply glue to the joint. If the wood is to be stained, it's a good idea to add food coloring to the glue. That way, any squeeze-out left on the wood after wiping with a damp rag will be easy to spot. You don't want glue to dry on the exposed surface of stain-grade molding; it repels stain and leaves an unsightly discolored spot.

Cutting and Fitting a Butted Frame

The first trim to be installed is the head casing, the same as the mitered frame. If the head will receive additional moldings, creating a *cabinet head*, fasten those moldings to the head casing before installation.

In a butted frame, the head casing usually extends past the outer edge of the side casings, and for good reason. If the side moldings expand or contract, it won't be apparent. It's almost impossible to see slight variations in the head casing overhang. But if the head is flush to the edge of the side casings, any change in the side casings will be apparent. The tops of the side casings won't be flush with the top of the head casing if any movement occurs. I usually allow a 3/16-inch overhang at either end. See Figure 8-1.

Draw reveal lines on the jamb the same as for mitered casings. Install the head casing flush to the reveal line, making sure the overhang extends equally to either side. Hold the side casings in position and mark their intersection with the head casing. Cut the side casings to length and fasten them in place. I recommend cutting these a little long (about 1/32 inch) if they run to the finished floor. They should bow out from the wall about 1 inch at the most when fitted into place. Nailing the bow to the wall ensures a tight joint.

Block and Plinth Door Casing

Start by drawing reveal lines on the jamb the same as for mitered and butt joint casings. Then cut the header casing to length. The blocks will be located at the intersections of the side reveal lines and the header reveal line. See Figure 8-6. Cut the header casing to fit the space between these two points.

Next install the plinths on the finish floor, or on blocks representing the thickness of the finish floor to be installed. Square-cut the bottom of a side casing and set it on a plinth. Mark where the top edge meets the bottom corner of the header casing. Cut and install the side casing to the reveal line. Repeat the process for the second side casing. Finally, install the corner block in the square space where the head case meets the side trim. The block itself probably isn't perfectly square, so plane it to fit snugly.

Manhours for Installing Trim

All figures in the tables are in manhours and are based on the following assumptions:

- Tools and materials needed are available on site.
- The tradesman is a qualified and motivated finish carpenter.
- Work is good quality, stain grade, done no more than 9 feet above floor level.
- All defects are remedied before the carpenter leaves the site.

Add extra time for setup, cleanup, painting or staining, protecting adjacent surfaces, complicated layout or inadequate plans, repair and replacement jobs where fitting and matching is required, working around other trades, setting up scaffolding and ladders for work above 9 feet, and supervision, if necessary. Paint-grade work will usually reduce the time needed by from 20% to 33%.

Set Casing Around a Framed Opening

Mitered or butted molding corners, per opening, using a pneumatic nail gun. Add 50% for casing over 2¼" wide or for hand nailing.

Opening to 4' wide using casing up to 2" wide	.30
Opening over 4' wide to 8' wide, using casing up to 2" wide	.50
Add for casing over 2¼" wide to 4" wide, per opening	.16
Add for block and plinth door casing, per opening	.30

Chapter 9

Window Trim

It's just plain luck if we finish carpenters are on site when the windows are installed. Nine jobs out of ten you won't see the windows being installed, which probably means you're going to have problems with the window trim. So before you start trimming out any windows, check how each unit was set in the wall frame. Use Figure 9-1 as your checklist.

Unless otherwise specified, use the same style of casing on the windows as you used on the doors. There are two basic styles of window casing:

1) The mitered surround case frame, also called a *picture frame*. See Figure 9-2. This type is usually mitered at all four corners. Occasionally, corner blocks are used.

2) The stool and apron case frame. It may have corner blocks, as shown in Figure 9-3. It may also be mitered to the head casing.

Some contemporary window installations have no jambs or casings at all. The window opening is simply wrapped with drywall. This style usually includes a stool supported by an apron.

Preparing for Trimwork

Prep work on window cases is basically the same as casing on door jambs. Make sure you install the window cases in proper construction sequence, as mentioned in the last chapter.

Inspection Checklist for Window Installations

☐ Like-size windows should be set to a uniform height throughout the building. Paired windows must be set precisely so the head casing and stool match perfectly.

☐ Window jambs should be square. Check squareness with pinch sticks as in Figure 6-8 in Chapter 6. Poorly installed double-hung windows are easy to spot because the mid-sash of each isn't aligned.

☐ The side jambs must either sit directly on the frame sill, or on shims. Wide windows, especially those with mullions, are supported under their sills as well.

☐ The jambs should be flush to the wall surface unless extension jambs are specified.

☐ The jambs should run straight. Look out for bows caused by insulation stuffed or injected between the jambs and the frame.

☐ Be sure the drywall doesn't protrude out from the window jambs.

☐ Check the drywall to be sure it's securely fastened in the trim areas.

Figure 9-1 **Checklist for window trim**

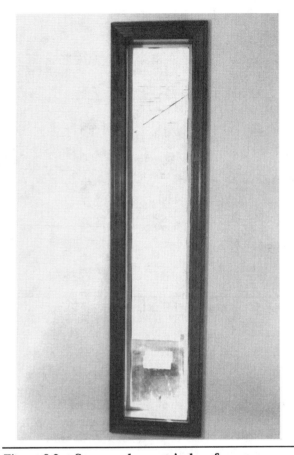

Figure 9-2   **Surround case window frame**

Photo by Pat Cudahy

Figure 9-3   **Stool and apron window case frame**

If floor-to-ceiling paneling will be installed, the time to do it is before you begin on window trim. Cut extension jambs, if needed, and fit them to the window before starting on the casing.

Many newer homes have 2 x 6 stud walls instead of 2 x 4 stud walls. The extra space is needed for insulation in the walls. If windows designed for a 4-inch stud wall have been set in a 6-inch stud wall, you'll have to install extension jambs. *Extension jambs* bring the window units out flush to the interior finished wall surface.

If you're lucky, extension jamb stock will be supplied with the window units. You may, however, have to mill this trim yourself. Begin by cutting the extension stock to rough length. Leave 1 inch for waste on the ends and ½ inch extra in width. Leave these trim sets bundled by each window. When the stock is laid out for all the windows, you can go back to the first window and start installation.

If the casing style is a full surround, you'll need to cut extension jambs against the sill as well as against the two sides and the header. For a stool and apron window, cut only side and header extensions. The stool is installed along with the extension jambs. Let's take a closer look at working with each of these styles.

The Full Surround

Begin installing extension jambs by laying out the reveal lines on the edges of the window jambs. Use the same wood layout block I described in Chapter 8.

Some finish carpenters don't leave a reveal where the extensions meet the window jambs to avoid creating a shadow line. This method is perfectly acceptable, but it's much harder to cut and install. Also, uneven shrinkage or expansion of the jamb and extensions will create a shadow line here anyway.

With the reveal lines marked, determine the finished lengths of the extension jambs and cut them to size. I usually butt the sill and header into

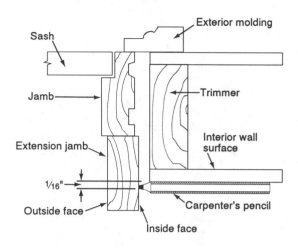

Figure 9-5 Window trim (plan view)

Figure 9-4 **Extension jamb box held temporarily with shims**

the sides. Rabbeting isn't necessary, in my opinion, so I don't account for it in my measurements.

Now assemble the extension jambs with screws and hold them temporarily in place with shims wedged between the corners of the box and the stud frame. See Figure 9-4. Check your reveals for accuracy. If everything looks okay, mark the width of the extension jambs by running a pencil line down the inside face of the extensions. See Figure 9-5. Note that the line created by the pencil is ¹⁄₁₆-inch beyond the actual flush line. This leaves room for planing in a back bevel as described below. Be sure to use a sharp pencil when marking this line.

After marking the perimeter of the box, remove it from the opening. To make sure the jamb extension gets reassembled and reinstalled the same way, mark the inside of the four edges, as shown in Figure 9-6. Notice that the four marks are four parts of a pyramid: the base goes on the bottom extension, the cap on the top and the left and right parts of the pyramid go on the left and right extensions.

Now take the box apart and cut the components to width. With a small circular saw and small diameter blade, cut to the outside of the pencil line. To cut along a curved line, set the blade to the thickness of the stock plus the depth of the tooth gullets. Use a hand plane to smooth the edge and create a back bevel in the edge. A bevel of 2 to 3 degrees away from the outside face helps fit the casings snug against the edge.

Reassemble the extension jamb box. For quality workmanship, use spline biscuits inside the butt joint in addition to screwing and gluing.

Now place the box back in the opening and insert sets of shims between it and the stud frame. Bring the extension jambs even with the reveal

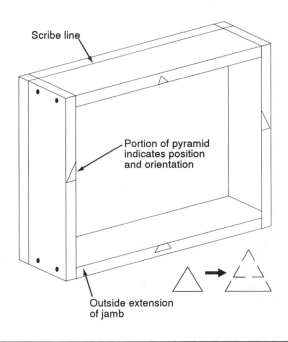

Figure 9-6   **"Pyramid" marking method**

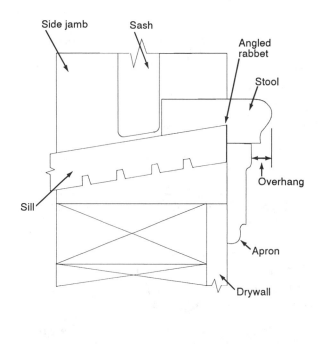

Figure 9-7   **Window trim cross-section**

lines by adjusting the shims. When everything lines up, hold the box tight against the window unit and nail it to the frame through the shim sets.

Installing Extension Jambs with Stools

As I mentioned before, attach the stool to the extension jambs before scribing and installation. First, though, cut the stool to fit the opening and the window frame.

Stool stock is available in a variety of widths. It's usually sold with an angled rabbet precut in the bottom to clear the inside corner of the window sill. Figure 9-7 shows the stool rabbeted to fit over the angled sill. The stool should overhang the apron by at least ½ inch. If you're installing to metal window frames, you won't need a rabbeted stool.

Sometimes the stool stock available isn't wide enough to both reach the sash and provide enough of an overhang beyond the apron. In this case, glue an extension to the inner edge of the stool molding. Be careful not to create a visible reveal at this joint. Glue this extension with water-resistant carpenter's aliphatic resin glue. Your local building supply store probably offers this glue.

Begin the fitting process by cutting the stool to the right length. Find this length by laying out the widths of the casings on the wall the way they'll be when they're installed. Measure the span between the outside lines. To this add the sum of the overhangs created where the stool extends beyond each side casing. The extension should be equal to the distance the stool sits out from the face of the casing. The line of the

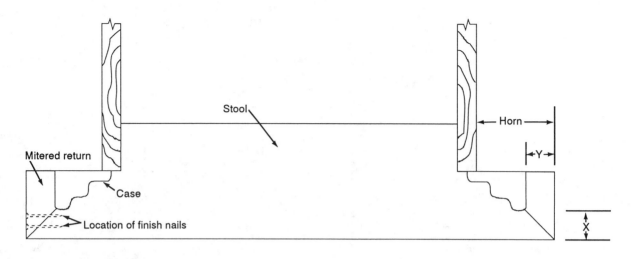

Figure 9-8   **Finished window stool (plan view)**

mitered end return will then pass through the corner of the case molding. That makes for a very professional-looking joint.

Figure 9-8 shows a plan view of the finished window stool. Notice the mitered return of the sill horn, the portion of the stool that goes past the outside edge of the side casings. Overhang Y has to equal overhang X if the return miter is going to intersect the corner of the casing.

Now miter the ends of the stool to this length and glue the return miter pieces into place. Since you still have to cut the horn, be sure to position the finish nails out of the line of the horn's rip cut. See the left side of Figure 9-8. These nails hold the returns in place while the glue sets. When dry, take a little time to apply primer or stain and sealer to the underside of the stool.

To cut the horns, hold the stool against the window, the ends to their marks. Now determine the depth of the horn by measuring the distance the stool must move to hit against the window sash. That's distance X in Figure 9-9. Slide a combination square along the outer edge of the stool until the square hits the corner of the jamb. Draw a line along the arm of the square. Mark distance X on this line, being sure to measure from the edge of the stool facing the window sash. Repeat this procedure for the horn on the left.

Since the wall may not be perfectly parallel to the front edge of the stool, I recommend that you set a scribe to the distance between the intersection mark and the corner of the window jamb, then run the scribe along the wall to the end of the stool. To make an accurate line, it's essential that you hold the scribe perpendicular to the stool's front edge as it traces wall irregularities. See Figure 9-10.

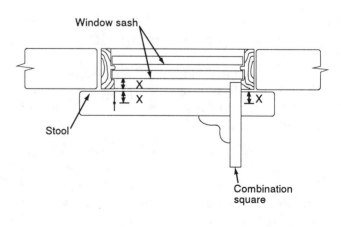

Figure 9-9   **Transferring a jamb corner to a stool**

Figure 9-10 **Using a scribe to mark horn cut**

Figure 9-11 **Installing a stool and joint assembly**

Cut the horn at each end of the stool stock on the lines measured with the combination square and the scribe.

When you're satisfied with the fit of the stool against the wall and against the window sash, fasten the jamb extensions to it and fit the assembly into the opening for scribing. Follow the same process for fitting and installing full surround extension jambs. Figure 9-11 shows a stool and joint assembly fitting into the opening.

Installing Stools in a Jambless Opening

If the style of the window trim calls for only a stool and apron without jambs or casing, then you find the crosscut for the horn in an entirely different way. Figure 9-12 shows the sequence of steps for this. If the stool is to be fitted to a preinstalled exposed side jamb trim or to a finished drywall surface, this sequence is necessary for finding the proper angle of the horn's crosscut.

To begin, support the stool on a couple of sticks temporarily fastened to the sill. Then use a combination square to determine the point where the rip line meets the crosscut line. This is illustrated in Figure 9-12A. Ignore the square line itself. Replace it with a line established by running a bevel gauge along the front edge of the stool. Have the arm set against the inside drywall wrapping. See Figure 9-12B. Note that any angle other than 90 degrees brings the bevel gauge away from the intersection point. Slide the stool toward the bevel gauge to draw the line along the blade to the intersection point. See Figure 9-12C. Use the scribe to determine the rip cut of the horn.

Cut the horns with a jig saw. After testing and trimming with a chisel, check to see if they fit in the openings. The stool should be snug to the

A Step 1

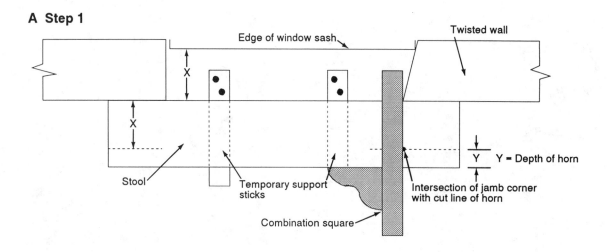

B Step 2

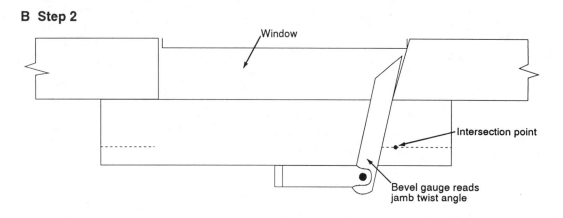

C Step 3

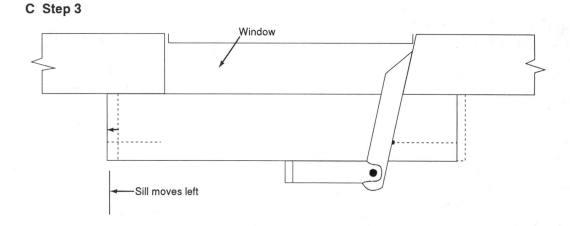

Figure 9-12   **Marking a horn on a jambless window opening**

Lift the assembled frame to the window and install it. The frame will be flexible enough to allow some adjustment if you're out of square.

Casing Surround Frames: The Surround Method

If the window jamb is more than 3/16 inch out of square (the difference in diagonal measurements), it's best to cut and fit the surround frame in place — one piece at a time.

Start with the head casing. Cut it to length with a 45-degree miter on each end. Install this head. Now fit the right-hand side case. First cut a 45-degree miter on one end of an oversize length and try the fit. A small amount of planing may be necessary to change the miter angle until the side case runs parallel to its reveal line.

When you're satisfied with the joint, cut it to length with a 45-degree miter, starting at the bottom end at the mark. Nail it in place along its length. Don't forget to glue and nail across the upper miter.

Install the bottom piece the same way. But be sure to leave the last 12 inches of the left-hand end free of nails. This lets you shift the molding a bit, making the last joint a little easier to fit.

Find the length of the left side case molding by holding the outside edge of the rough length of stock against the points of the two horizontal moldings already installed. Mark the location of the miter points. Now make a 45-degree miter cut on the molding at the marks, plus about 1/8 inch for final planing to fit. When you're satisfied with the fit at top and bottom, glue the joints and nail the molding.

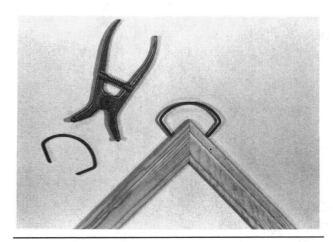

Figure 9-13 **Holding a miter joint with a spring clamp**

window frame. The horns should fit tight against the walls. If necessary, shim the stool level. Finally, secure it with finish nails.

Casing Surround Frames: The Picture Frame Method

Let's assume you're going to do a full surround case frame with mitered corners. First, select stock for the frames. If you're using stain grade stock, take all casings from a single piece so the grain matches. Cut the casings' ends to 45-degree angles and at the proper lengths. It's easy to find casing lengths from the reveal lines you've already made on the jamb edge: outside length equals the distance between the reveal lines plus twice the width of the molding.

When the casings are cut, assemble the frame on a flat work surface, not around the window itself. Slot for and install spline biscuits into the miter joints and glue the corner joints together.

A spring clamp is useful for holding the joint tight while the glue dries. In Figure 9-13, note the pliers-like tool used to install and remove the springs. A miter clamp along with spline biscuit joinery almost guarantees these miter joints will be permanently gap-free.

Stool and Apron Frames

The stool, but not the apron, should already be installed before you begin installing the casing for a stool and apron frame.

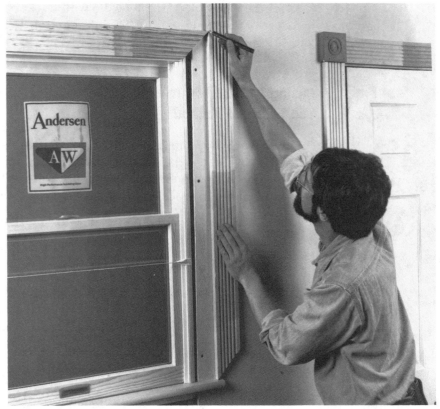

Photo by Pat Cudahy

Figure 9-14 **Edge marking a header miter**

Installing the Casings

Install the header casing first. Then fit the sides in between the head casing and the stool and mark the length of the side cases:

1) Cut a miter in one end of a rough length.

2) Stand the casing on end so that the point of the miter is facing down and resting on the stool.

3) Now mark the point of the header miter on the side of the side casing. See Figure 9-14.

4) Install the side casings to the wall and jamb. Then secure the stool to the bottom of the moldings by nailing up from below.

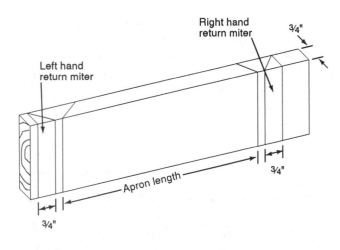

Figure 9-15 **Laying out return miters**

Installing the Apron

Generally, the apron molding should be cut from the same material as the side and head casings.

Cut the length of the apron to equal the outside span of the side casings. Shape the ends to the profile of the molding. This is easiest done by return mitering the ends. Figure 9-15 shows the sequence of crosscuts made in a rough length of apron stock. These cuts produce the apron and provide it with a return miter at either end.

If the molding profile has a rounded corner where it meets the underside of the stool, rip or plane it off so it fits without a visible gap.

Install the aprons without the returns in place. After cutting the apron to length with miters at either end, glue along the top edge where it meets the underside of the stool. Gluing the apron to the sill instead of nailing it eliminates unsightly nail holes along the top face of the stool.

Cutting and Installing Molding on a Curve or Arch

Arched windows usually require special moldings, and special techniques to cut and fit them in place. I'll use a simple arched window frame to illustrate how to make up the arched head casing, and how to determine the angle of the joint with the side casing.

Determining the shape of the arch- The first step is to create a full-size pattern of the arch. Make the pattern by tacking a piece of building paper or asphalt felt over the installed window jamb. The paper must be at least the size of the opening plus the width of the molding. Rub the outline of the inside corner of the jamb into the paper with the edge of a carpenter's pencil. Now remove the paper from the opening and lay it out on a flat surface.

Add the width of the desired amount of reveal to the rubbed line. This establishes the inside line of the arched molding. Establish the outer line by setting your pencil scribe to the width of the molding. To create an accurate pattern, hold the scribe parallel to the radius of the arch at all points along its inside line.

Figure 9-17 shows a typical arch. Note how this arch is made by the joining of three circles. If you hold the scribe parallel to the radius of each circle, it transitions smoothly from one circumference to the next.

If the arch falls on the circumference of a single circle rather than a combination of three circles, use trammel points mounted on a stick.

Photo by Pat Cudahy

Figure 9-16   **Using a springboard to hold apron to stool**

Clamp the molding in place, using a springboard to hold the apron to the stool. See Figure 9-16. Make sure the ends are flush with the outside edge of the side casings resting on top of the stool. Fasten the apron to the wall with a pair of nails into the trimmers at either end. Drive several single nails in between into the frame sill. Apply glue to the return miter pieces and press them into place. If pressure between the apron and the wall isn't enough to hold them in place while they dry, use spring clamps.

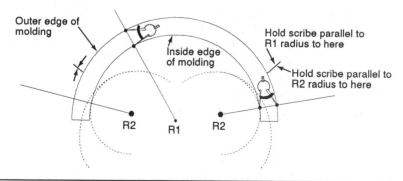

Figure 9-17   **Scribing outer radius of a curved molding**

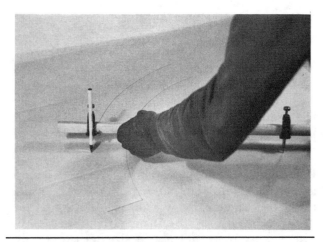

Figure 9-18   Using trammel points

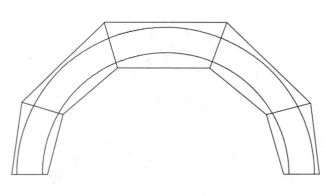

Figure 9-19   Board layout for arched molding

See Figure 9-18. First, find the center of the circle. Then set the trammel point with the pencil inserted at the desired outside radius of the arch. Draw the outside edge of the molding as shown in Figure 9-18.

Laying out the boards that make up the arch- With the pattern of the arched molding drawn full scale on paper, it's easy to lay out the stock from which the molding will be cut. Keep the width of the boards needed to furnish each portion of the curve as narrow as possible in order to help keep the grain parallel to the curve. If the arch will be stained, try to cut all material from the same board. Figure 9-19 shows the optimal layout for this particular radius of arch.

Joining the components of the arch- Lay your drawing down on a flat surface and determine how many pieces of what length boards are needed. Take angle and length measurements for the end cuts directly from the drawing. Then miter-cut the pieces that will make up the arch. To check the accuracy of your work, cut out the paper pattern of the arch. Lay this pattern over the arch pieces. If the arch pieces extend beyond the edge of the paper pattern at all points, begin assembling the arch.

Join the arch pieces with spline biscuit joints or dowels. Don't use nails. Any nail driven in the arch will almost certainly end up where you want

to make a cut. Assemble the frame on a flat surface where blocks and wedges can be used to hold the arch securely while the glue sets.

Shaping the arch- When the glue has dried completely, stick the pattern to the arch with double-stick tape. Use a sharp pencil to mark the pattern on the boards. Then remove the paper and cut out the arch with a jig saw or band saw.

Smooth the edges of the arch with rasps and cabinet scrapers. Then cut two sanding blocks. One should be cut to the shape of the inside radius and one should be cut to the shape of the outside radius. Sand with these blocks until the arch is smooth. If the edges are to receive a profile, this is done either with a shaper or a router set up with a ball-bearing pilot on the cutting bit.

Finding the angle between the arch and the side casings- Usually you'll have to join the arch to some other molding. If the arch is a perfect half circle and joins casing at both ends, you can make a square cut at the top of the side casings and butt the arch to those ends. If the arch isn't a perfect half circle, you'll have to make a miter joint.

Here's how to determine the angle of this miter joint. Follow along on Figure 9-20. First, hold the side casings in the correct position on the wall. Be sure the inside edge is along the

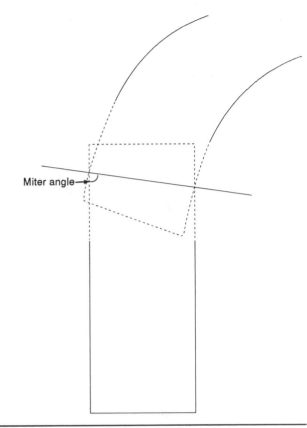

Miter angle

Figure 9-20   **Miter joints in curved molding**

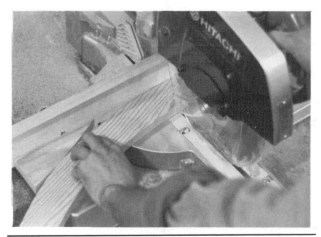

Figure 9-21   **Using a chop saw with a backing block to cut an arched molding**

reveal line. Trace the outline of the top of the casing onto the wall with a sharp pencil. Remove the moldings from the wall.

Second, hold the arched head casing in place (it's easier if you temporarily secure it with a couple of unset finish nails) and trace the outline of the two ends. Since you cut the arch and the side casings overlength, the lines will intersect each other as illustrated in Figure 9-20.

Third, remove the arch, then draw a line connecting the two intersection points. This line represents the miter angle you'll cut to join the two moldings together.

Finally, hold the arch back in position and mark the intersection points on the edges of the molding. You'll use these marks to orient the piece in the chop saw. Use the bevel gauge to set the chop saw for cutting the angle into the side casings.

Cutting the angles- Figure 9-21 shows a chop saw cutting the miter angle into the arched head casing. If the arch is large, get a helper or use a support stand to hold the molding level while you cut it. Insert a block between the molding and the chop saw fence to add support backing and to hold the proper angle of the cut. Make several test cuts first in the waste area before committing yourself.

Installing the arched molding- When the casing and arch have been miter-cut to length, install them to their reveal lines. Begin with the side casings, but don't nail the top 12 inches of casing so you can make adjustments, if needed. Then set the arch in place on top of the side casings. When positioned properly, nail all molding to the jambs and trimmers. Do a little smoothing and shaping at the joint where the arch meets the side casings. Use a carver's palm chisel and sandpaper wrapped around shaped blocks. Figure 11-9 in Chapter 11 shows how to use sanding blocks.

Manhours for Installing Window Trim

All figures in the tables are in manhours and are based on the following assumptions:

- Tools and materials needed are available on site.
- The tradesman is a qualified and motivated finish carpenter.
- Work is good quality, stain grade, done no more than 9 feet above floor level.
- All defects are remedied before the carpenter leaves the site.

Add extra time for setup, cleanup, painting or staining, protecting adjacent surfaces, complicated layout or inadequate plans, repair and replacement jobs where fitting and matching is required, working around other trades, setting up scaffolding and ladders for work above 9 feet, and supervision, if necessary. Paint-grade work will usually reduce the time needed by from 20% to 33%.

Setting Window Trim

Based on trim set around prepared openings, per window, based on square windows. Extra tall or extra wide windows may take longer.

Extension jambs cut, assembled, shimmed and installed

Up to ¾" thick, 3 or 4 sides, per opening	.33
Over ¾" thick, per linear foot of extension jamb	.10

Window stool

Set to sheetrock wraps, nailed directly to the framing, per opening	.50
Add per linear foot of stool when stool has to be shimmed up from framing	.05
Set to wooden jambs, per opening	.40
Add per linear foot of stool when stool has to be shimmed up from jambs	.05
Add for mitered returns on horns, per opening	.50
Window apron molding without mitered returns, per opening	.16
Window apron molding with mitered returns, per opening	.32

Window casing, straight casing with square corners, per opening

3 sides (stool and apron), up to 5' on any side	.33
4 sides (picture frame), up to 5' on any side	.50
Add for any side over 5' long, per side	.09
Add for angle corners (trapezoid or triangle), per corner	.25

Installing an arched top casing only, assuming a premanufactured arch

To 4' wide set against rosettes	.33
To 4' wide, mitered corners	.50
Over 4' wide set against rosettes	.66
Over 4' wide, mitered corners	1.00

Chapter 10

Ceiling Treatments

In this chapter we'll take a look at installing several types of ceiling treatment. Although cornice molding is really a type of ceiling treatment, I'm going to wait and deal with that one in the next chapter.

As I've mentioned before, finish carpentry is best done from the top down. Work on ceilings should be done before other work in a room. That reduces the risk of damage to work you've already finished.

Before you begin, be sure the ceiling is properly prepared. Check to see that:

1) The drywall is securely hung, taped, and sanded smooth.

2) Furring strips are installed, if needed.

3) The ceiling is adequately level and flat.

Anything that penetrates the ceiling and requires trim work cut out around it should be in place — skylights, chimney frames, chasers for utilities, and the like. Cornice soffits also need to be in place before you begin.

Scaffolding

Unless the ceiling work is really simple, like installing a few applied cross beams, you're going to need scaffolding. Trying to make do with stepladders won't work. It's not only a waste of time, it's just plain not safe. Don't discover that the hard way.

Scaffolding should reach the perimeter of the room and be high enough so you can place your outstretched palm flat to the ceiling.

Consider placing your cutting station on a centrally-located scaffold, especially if you're working alone or if you need a lot of assistance from a helper. This way you won't have to climb up and down a ladder to make every cut. Even if the station needs to be relocated as the work progresses, a cutting station on the scaffold makes sense. Relocation is a small inconvenience compared to the time and energy it saves.

Safety tip: Always inspect your scaffold thoroughly before using it. Be sure all diagonal struts are in place and that the connections are secure. See that the legs sit firmly on the floor and the casters lock properly. Make sure that blocking under legs that span openings in the floor (such as stairwells) is well secured. Then inspect the scaffold boards. Look for the stamps that say *scaffold grade*. Check for signs of splitting or rot. Refuse ungraded stock with large knots or excessive grain run-out. Sight along the edge to check for grain that's more than 20 to 30 degrees out of alignment with the surface.

Figure 10-1 shows four kinds of defects common in scaffold plank:

1) Unusually wide growth rings.

2) End and shake splits.

3) Large centrally-located knot.

4) Cross grain that's at a large angle to the face.

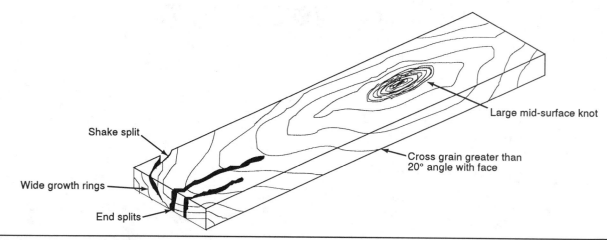

Figure 10-1   **Scaffold defects**

Installing Strip Ceilings

One common type of ceiling treatment is the wood strip ceiling. The strips are usually of 4/4 (¾-inch) stock and less than 6 inches wide. They're either tongued and grooved or have square edges. The square-edged strips are spaced slightly apart. Generally square-edged strips are applied over a ceiling painted a subdued color.

Many finish carpenters, myself included, prefer their stock prefinished. If the ceiling strips aren't prefinished, at least make sure that the back of the wood (the side facing the ceiling) is primed or sealed before installation. Otherwise, the wood may eventually warp due to uneven moisture absorption.

Begin by snapping chalk lines on the ceiling to indicate the position of joists. Then measure out evenly from the opposite wall and snap guidelines across the ceiling about every 6 feet. These lines will run parallel to the strips, making an easy reference for keeping the ceiling strips parallel to the opposite wall.

When layout is complete, you're ready to install the first run of stripping. With tongue and groove stripping, face the tongue out, toward the center of the room, and nail it off. Use a pair of nails at each joist; one through the face and one through the tongue. The rest of the runs (except

the closing strip) need tongue nails only. With non-tongue and groove strips you'll have to face nail each run.

Be careful to keep a perfectly straight edge line when you're butting lengths of starting strip end-to-end. Check the line by stretching a string from one end of the run to the other. Make sure you join the butts over a joist so you get adequate nailing on the ends.

As you install the strips across the ceiling, keep measuring from the edge of the runs to the chalk lines to make sure the runs are staying parallel. Make small corrections by adjusting the amount of gap between the strips. If the runs are way off, make corrections in small increments over the next three to four strips. That way the adjustment won't be as obvious.

Here are a couple of simple tricks that will ensure a good fit between the runs:

1) If the strips are to be spaced, set a temporary nail the diameter of the desired gap into the joist against the front edge of the last installed strip. See Figure 10-2. Draw the next strip to the nail and face nail it. Remove the spacer nail, reset it to the new edge face and move on to the next joist.

Figure 10-2   **Using a nail as a spacer between ceiling strips**

Figure 10-3   **Using a pry bar with a protector block**

2) Draw tongue and groove strips tightly together with a small pry bar and a protector block. Make the block from a 4- to 6-inch length of tongue and groove stripping ripped in half. Figure 10-3 shows how it's inserted over the tongue while the pry bar levers against it, pulling the strips together.

Note also in Figure 10-3 the nail being driven through the tongue into the joist above. If air-driven nailing isn't available, then use square flooring nails. They're less likely to split out the tongues than are common rounded-shank nails. With rounded-shank nails, it's best to drill a pilot hole.

The last strip can be tricky. There's almost always a taper or a curve that must be ripped into the edge of the strip that meets the wall. To get this line, measure the width of the gap between the last ceiling row and the wall at each joist location. Now transfer these measurements to the board to be cut. Lay out the spacing of the joists on the board and mark the width of the gap at each station. Call out the numbers to a helper or make a sketch of the closing strip and carry it to the strip to be laid out and cut.

Figure 10-4 shows a cutting pattern sketch for the closing board of a strip ceiling. The bottom numbers refer to the "stations" measured from the left end of the gap to be filled. The top numbers indicate the width of the gap measured at that station.

Connect the marks made at each station. You can use a straightedge or a batten, but a $\frac{3}{8}$ x $\frac{3}{8}$-inch strip of wood does a better job. See Figure 10-5. It curves as it's held to each mark, so it's more likely than a straightedge to duplicate the shape of the gap.

Cut the closing board with a jig saw or small-bladed circular saw. Back-bevel the cut 2 or 3 degrees so the board clears the wall and swings easily into place. Without a back bevel, the bottom of the cut hits the wall before the top

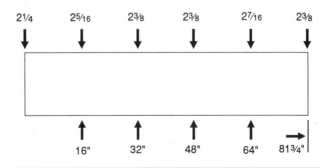

Figure 10-4   **Sketch of closing strip**

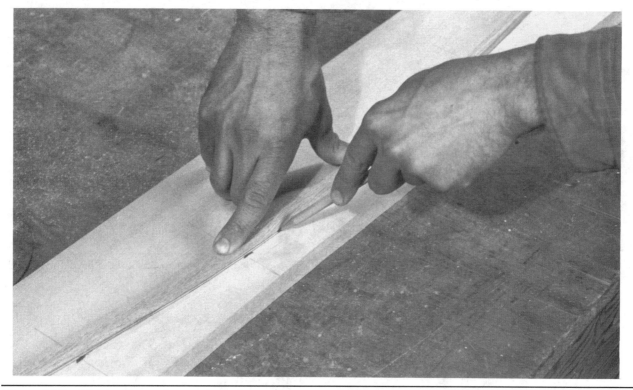

Figure 10-5 **Using a batten to connect the marks**

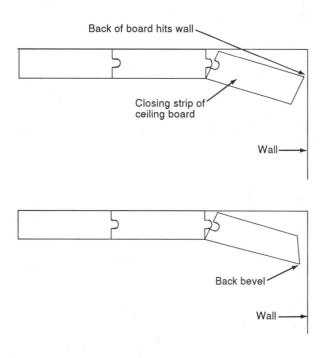

Figure 10-6 **Back-beveled closing strip**

does. See Figure 10-6. Now hammer the strip into the gap. Be sure you protect the finish surface with a scrap of wood.

Installing Panel Ceilings

There are basically two kinds of ceiling panels: solid stock frame and panel units and plywood sheet paneling.

Solid stock frames and panels are usually prefabricated and tend to be heavy. Make sure the components aren't too big. Large panels are awkward to install and join properly.

A panel treatment generally requires furring on the ceiling. As mentioned earlier, furring makes it easy to level the ceiling. It also allows you to place solid backing at the joint between panels. Run the strips on 16-inch centers perpendicular to the joists.

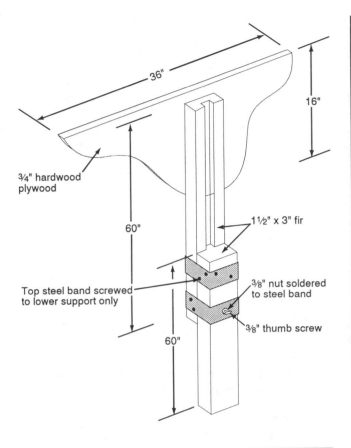
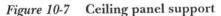
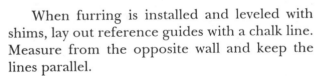

36"

16"

¾" hardwood
plywood

60"

1½" x 3" fir

Top steel band screwed
to lower support only

⅜" nut soldered
to steel band

⅜" thumb screw

60"

Figure 10-7   Ceiling panel support

Figure 10-8   Drill press with plug cutting bit

When furring is installed and leveled with shims, lay out reference guides with a chalk line. Measure from the opposite wall and keep the lines parallel.

Begin installation with a corner panel. Use a shop-made panel support as shown in Figure 10-7. Unlike most commercial lifts, it works as well on a scaffold as on the floor. The two 1½ x 3-inch legs are fitted with steel bands. Note the band with the nut soldered over a hole. A thumb screw, passing through the hole, locks the support anywhere from about 76 to 100 inches. The tongue and groove between the legs is optional, though it does add extra rigidity and makes for easier operation. Keep it waxed with paraffin from the stub of a candle.

Once the panel is in place, nail it off and remove the support. Sheet paneling ¼-inch thick should be nailed at 8- to 12-inch intervals along the run of the furring strips. The nails will be visible, so use threaded panel nails with colored heads. Along edges that will be covered with molding strips, use either air-driven finish nails, staples, or drywall screws.

Attach heavier ¾-inch solid wood frame and panel units with screws driven through furring strips into the joists. Unless the panels are MDF or plywood, sink screws only through the frames. This gives the panels freedom to move in their frames.

Drill pilot holes for the screws and include a ⅜-inch countersink hole. Cut bungs from the same wood as the panel stock and glue these bungs into the countersink hole. Chisel off any protrusions and then sand flush. Figure 10-8 shows a drill press with a plug cutting bit that creates the bungs. Most mail-order tool supply companies sell plug cutters and drill bits with countersinks.

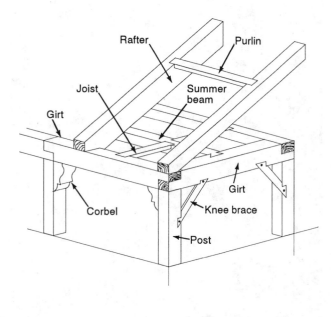

Figure 10-9 **Post and beam elements**

Let's begin by defining the basic parts (see Figure 10-9):

1) The *posts* are the vertical members of the frame. They support the horizontal members of the frame, called *beams*.

2) Beams either sit on top of posts or, in multiple-story frames, are joined into the posts.

3) Secondary beams, referred to either as *joists* if they're in a floor or *purlins* if they're in a roof frame, are often joined into larger beams. They're never supported by posts.

4) The larger primary beam is called a *rafter* if it's in an angled ceiling, a *summer beam* if it's in a flat ceiling, or a *girt* if it joins the outside corners of a room.

5) *Corbels* refer to the solid brackets often placed where beams join posts. Corbels add compressive strength to the posts.

6) *Knee braces* are diagonal members that stand away from these joints and add diagonal strength to the frame.

Figure 10-9 shows the basic elements of post and beam construction and the relative proportion of the members of the frame to each other. Note the orientation of the summer beam to the pitch of the roof.

Once you've got the corner panel installed, install the remainder of the first row of panels. Then proceed with the next row toward the opposing wall. Cut and fit the last row by measuring widths at each joist location, about like you fit the closing board on a strip ceiling.

Applied Beams, Posts and Corbels

Imitation beams, posts, and corbels applied to the interior of a room are intended to create the appearance of a post and beam frame. Creating the illusion of traditional post and beam construction requires some understanding of that type of construction. I'll stop for a moment here so we can take a quick look at this type of framing.

Building Up Posts and Beams from ¾-inch Stock

Unlike the real thing, the imitation post and beam framing usually isn't made of solid stock. There's no need to mortise and tenon the framing together because these "posts" and "beams" don't support any of the building's weight.

Instead, you create the look of solid timbers by joining together three-sided boxes made from ⁴/₄ stock. See Figure 10-10. Traditional timber framing was either oak or pine on the East Coast and fir on the West Coast. If your imitation posts

Post construction

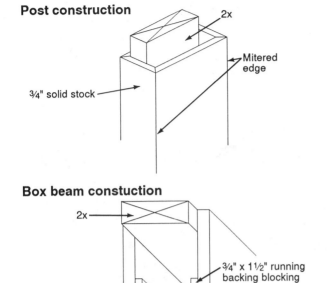

Box beam constuction

Figure 10-10 **Post and beam box construction**

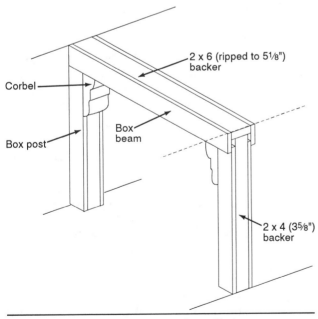

Figure 10-11 **Post and beam installation**

and beams will be painted, use the lightest material available, usually pine. The lighter weight makes installation easier.

Posts look best if the long edge joints of the box are made at a 45-degree angle. Done right, it's next to impossible to tell that the box isn't a solid timber.

Beams are done the same way. Or you can overlap the bottom board to create a pleasing shadow line. This makes cutting and installing a lot easier. Make the prefabricated boxes, or the box components, at least 1½ inches overlength so you can cut some off to make a precise fit.

Installing Post and Beam Boxes

Install three-sided boxes to backers fastened to the house framing. Make the backers from kiln-dried stud lumber (such as 2 x 6s) cut to the width of the inside dimension of the boxes. Try to lay out the posts so you can nail the backers to the vertical wall studs. Otherwise, you'll have to settle for nailing to the wall shoe and plate.

Run the beam backers perpendicular to the house joists or rafters. If the design calls for primary summer beams with secondary joists attached to them, be sure the backer for the summers runs perpendicular to the joists.

Install backers for the joist beams that run between the wall and the summer beams at the location of the existing ceiling joists. If you don't want to run your box beams at these locations, install girts to the outside corners of the room parallel to the summer beam. Fit the box joists between the summer and the girts and fasten them at either end.

Figure 10-11 shows a simple post and beam installation with corbels. Begin by installing a running backing block for the beam bottom board on the inside of the beam sides at the correct depth. (See Figure 10-10.) Then install the beam:

1) Plumb and nail the backers for the posts to the wall plate and shoe, and to a vertical stud, if present.

Figure 10-12 **Scribing beam side to wall**

2) Cut the backer for the cross beam to length. If it's two lengths, join the ends on a joist. Then nail the backer in place.

3) Cut one beam side 1 inch overlength and hold it in place against one wall. You'll have to hold it out from the opposite end of the backing. Scribe this end to the wall. See Figure 10-12. Cut to the line with a jig saw. Back-bevel the cut except at the bottom where it's exposed to the room. If the end of the beam takes a molding such as quarter round, you won't have to be precise. Just cut the end square within ⅛ inch of the wall.

4) Determine the total length of the beam side along the top edge, using your tape measure. Don't rely on the old rule of adding the size of the tape box to the last number you can read against the edge of the box and the wall. Instead, run the tape out from the wall to the right and make a mark at a whole number like 100 inches. Now run the tape to the left wall and add the number of inches that reach your 100-inch mark. Mark the length on the beam side along the top edge where the new cut will be made.

5) Hold the beam side in position to the opposite wall. Set the scribe to the distance between the upper corner of the wall and the overall length mark and trace the outline of the wall onto the beam side. Cut to this mark. Now install the side to the backer with finish nails.

6) Cut the other beam side to length and install it. Now cut the bottom to length and fit it between the two sides, nailing

Figure 10-13 **Routing out back of corbel**

it to the preinstalled running backing block. Scribing the ends isn't necessary as the posts will cover this joint.

7) Install the box post between the floor and the bottom of the beam. If the finish floor isn't in yet, you only need to worry about a gap-free joint at the junction with the beam.

8) Fit the corbels to the completed posts and beams. Plane the meeting surfaces of the corbel with a hand plane until you get a gap-free fit. Design the corbel so its width just slips between the side boards of the overlaying beam. If the corbels are fashioned from solid wood, use a router to remove most of the material facing the post and beam boxes. See Figure 10-13. Depth of removal need only be 3/16 inch. Hold it back about 1/2 inch from the exposed edges. This gives you less material to have to plane for a nice fit. Fasten the corbel into place with finish nails or screws and bungs.

Manhours for Installing Ceiling Treatments

All figures in the tables are in manhours and are based on the following assumptions:

- Tools and materials needed are available on site.
- The tradesman is a qualified and motivated finish carpenter.
- Work is good quality, stain grade, done no more than 9 feet above floor level.
- All defects are remedied before the carpenter leaves the site.

Add extra time for setup, cleanup, painting or staining, protecting adjacent surfaces, complicated layout or inadequate plans, repair and replacement jobs where fitting and matching is required, working around other trades, setting up scaffolding and ladders for work above 9 feet, and supervision, if necessary. Paint-grade work will usually reduce the time needed by from 20% to 33%.

Ceiling Treatments

All figures assume a wall height not over 9' with work done from scaffolding.

Set up and dismantle scaffolding, based on a typical 10' x 16' room, per room	1.75
Wood strip ceiling applied wall to wall with no miters, per square foot of ceiling	.10

Panel ceiling, plywood sheets and molding strips, including layout and installation.

Panels screwed in place, strips nailed with a pneumatic nailer, per square foot of ceiling	.16
Applied beams, including building up the beam, layout and installation on a flat ceiling but without scribing to the wall, per linear foot of beam	3.10
Add for each fitted joint or beam scribed to the wall	.66
Add for each corbel	.40
Add for each knee brace	.60

Chapter 11

Cornice Moldings

Cornice moldings are applied where the wall meets the ceiling. Styles of cornice molding vary. Some are plain, quarter round or flat molding similar to baseboard. The fancier ones have a special profile called a *crown*. Some trim in more traditional homes is highly decorative, such as crown and bed moldings run to a soffit box.

In this chapter, I'll keep it simple and just describe how to install a single run of crown molding.

Preparing for Trimwork

Though many finish carpenters don't bother, I think it's worth the time to put scaffolding around the perimeter of a room before you start installing a cornice. I find that it's faster and more convenient to work off scaffolding than it is to bounce up and down a ladder.

In large rooms, I also build a raised platform in the center of the room for the chop saw. I run at least two planks from the perimeter scaffolding over to the saw. Then I don't have to climb up and down the scaffolds to make each cut.

Remember my advice about working from the top down? Well, you should have the ceiling treatments in place before you begin installing the cornice molding. However, I would install full-height wall treatments and ceiling-height bookcases or cabinets before the cornice mold.

If the cornice will butt against chimney or utility chasers, complete the framing and drywall before you begin with the cornice molding.

Will the cornice molding be finished before you install it? Work that out with the painters. Personally, I prefer to install trim that's already finished. Then I go back later to fill holes and do any needed touch-ups. Either way, make sure the front and back of your cornice molding is at least primed or sealed before you start installation.

Installing Running Backing

If the cornice is wide crown molding or soffits, you'll need to install a running backing. For crown molding, use ¾-inch plywood or solid wood ripped to the angle the crown is designed to lay at against the wall — usually a 38- or 45-degree angle. For soffits, make the running backing from 2 x 4s. See Figure 11-1. Make sure your backings are level. You don't need backing on smaller crown molding because you nail it directly to the wall plates.

Lay out running backing, or narrow crown itself, to a level line run around the perimeter of the room (Figure 11-2). Start by establishing a starting point for this line:

- Find the low point of the ceiling with a long level and hold a sample of the soffit box or crown mold in position. Be sure the soffit sample is level, or that the crown molding is sitting at the proper angle.

- Mark the bottom edge of the molding or soffit box on the wall.

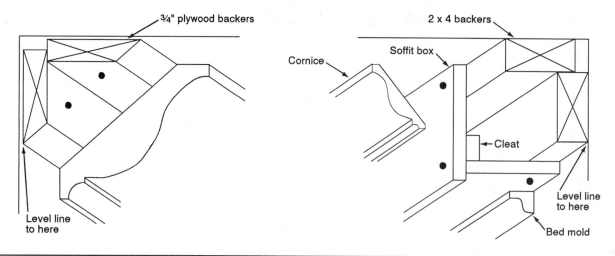

Figure 11-1   **Cornice running backing**

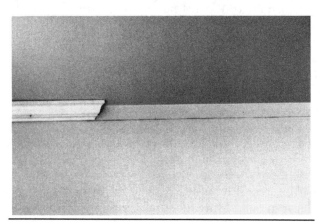

Figure 11-2   **Chalk line showing bottom edge of cornice molding**

- Measure up to the location of the lower edge of the running backing.
- From this point, draw a level line around the perimeter of the room. The easiest way is with a builder's level. If you don't have one, attach your most accurate level or an electronic level to a long straightedge.

If the ceiling isn't noticeably out of level, or if the cornice won't be running close to any level head casings, you may not need a level line. Instead, install the cornice parallel to the ceiling line. There's an advantage to this. The molding won't form gaps between its top edge and the ceiling, as leveled molding would.

Installing Typical Crown Molding Cornice

For the best results, plan the installation sequence. Which walls will be done first and which walls will be done last? A good sequence makes installation go smoothly and ensures efficient use of materials. And it minimizes the gap at joints if the molding should move or shrink. To plan the sequence correctly, you have to understand how joints are made in crown molding.

The Coped Joint

The inside corners of crown moldings and other moldings having a shaped profile are joined with coped joints rather than inside miters. While inside miters are easy to cut, they're difficult to fit. Unless the corner is perfectly square, you're going to end up doing a lot of planing and positioning to get a tight fit. Even then, the smallest movement in the molding is going to make an obvious gap between the mitered surfaces. I should warn you right here — no glue you apply to end grain can resist normal shrinkage and movement of wood molding.

A Unassembled coped cornice molding

B Assembled coped cornice molding

Figure 11-3   **Cornice molding**

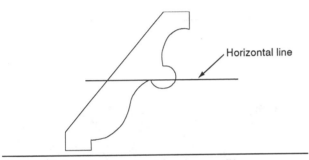

Figure 11-4   **Nonstandard crown profile**

The coped joint, on the other hand, generally keeps a good tight joint throughout the life of the molding. Here's why. The meeting edge of the coped molding is cut to a knife edge and pressure fit into the mating piece. Figure 11-3A shows an unassembled coped cornice molding. Figure 11-3B shows how it looks assembled.

I'll mention a caution here. You can't cope-join a nonstandard crown with a portion of its profile that dips below a horizontal line (as in Figure 11-4) without creating a gap. Try to avoid using crown moldings with this type of profile.

There are two ways to create a coped corner in standard crown molding.

Cope joining a crown molded inside corner-
1)   Fix a wood base and stop to the chop saw to hold the crown molding at the angle it'll take when you install it to the wall.

2)   Set the chop box to cut a 45-degree miter into the end of the molding. This creates a profile line which you can follow with a coping saw.

Note in Figure 11-5 that the molding is placed in the chop box upside down. The saw fence is in the position of the wall surface and the base is in the position of the ceiling. What will become the bottom edge of the molding is facing up.

3)   Now make the cut in the miter direction that makes this bottom edge of the molding the longest edge.

Figure 11-5   **Molding upside down in chop box**

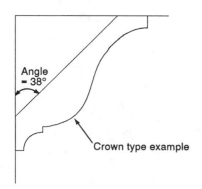

Angle = 38°

Crown type example

	Settings of compound miter saw to cut crown moldings sitting flat on saw table				
Type of crown molding	Cope on right end (Top edge of crown against saw fence)		Cope on left end (Bottom edge of crown against saw fence)		
	Miter angle	Bevel (tilt of blade)	Miter angle	Bevel	
45°	35.3° (right)	30°	35.3° (left)	30°	
38°	31.6° (right)	33.9°	31.6° (left)	33.9°	

Figure 11-6 **Angle settings for crown molding**

Alternative method-

1) If you're using a compound miter saw, you can cut crowns at the proper angle for coping with the molding laid flat on the saw base table. This is a great advantage in cutting wide moldings. It saves cutting by hand or on a larger specialty chop saw.

2) The trick here is to make the cut with the correct combination of miter angle and bevel. Figure 11-6 shows the settings for the two most common crown molding types. They're identified by the angle they make with the wall.

   Note: The orientation of the molding in relation to the fence depends on whether you make the cope at the right or left end. Also note that some compound miter saws have settings for these two standard cornice moldings. Test some cuts with waste stock to be sure the angles are correct for your molding.

3) Once the miter is cut, follow the profile line created by the miter cut intersecting the surface of the molding. Remove the waste with a coping saw, as shown in Figure 11-7. Make the cut with a

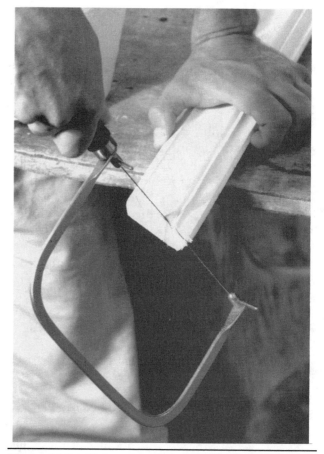

Figure 11-7 **Using a coping saw to remove waste**

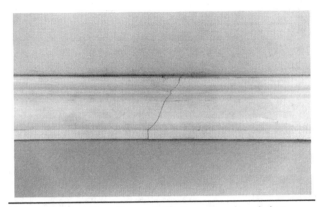

Figure 11-8   Overlapped 45-degree miter joint

considerable back bevel. Determine how much by holding the molding at the angle it's going to set against the wall. Now sight across the cut along the line the intersecting molding will take. There shouldn't be any extra material obstructing the back of the cut.

4)   Use round and triangular files to fine-shape the cut to make a perfect fit.

Making an Outside Miter Joint

Outside corners of cornice moldings are joined with a miter joint. To determine the location of this cut, hold one molding in place at the corner. Mark where the bottom of the molding passes the outside corner of the wall. The other end should already be cut to fit. It can either butt to a wall, be mitered, or be coped around an installed run of molding. The second piece of molding should butt to the wall at its opposite end. Since this butt end will be covered by a molding, you can make small adjustments at the joint without opening up the joint at the other end of the piece.

Make the miter joints on a standard chop saw, cutting a 45-degree angle away from the mark. The bottom of the molding is the short length. Note this is the opposite of the cut for a coped profile.

Place the molding on the saw table upside down and at the same angle it'll take when installed on the wall. The saw base is in the position of the ceiling and the fence is in the position of the wall.

If you're using a compound miter saw, lay the molding flat and set the angles as shown in Figure 11-6. Be sure the saw is cutting away from the mark. Again, the bottom edge of the molding is the short length.

Joining Lengths of Molding

When you have to join lengths of molding to make a long run, use an overlapping 45-degree miter joint. See Figure 11-8. This is called a *scarf joint.* Cut the miter in the underlying molding first. Then cut the overlying molding to length. Be sure you measure to the longest point. Cut a little long. Then trim to fit by nibbling back the cut with a chop saw or a block plane. Use a block plane if you're up on scaffolding and away from the cutting station.

Before nailing the lengths in place, apply a film of glue on both the outside miter joint and the mitered scarf joint. For a perfect fit, sand across the joint. Use a sanding block contoured to fit the molding profile, as shown in Figure 11-9.

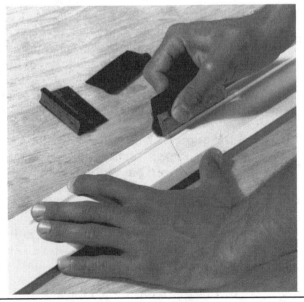

Figure 11-9   Using a sanding block

Sequencing Your Installation

Now that you understand how to make joints in crown molding, we can talk about the installation sequence. These are the general rules:

1) Make sure at least one part of an outside corner has a free end.

2) Place cope and scarf joints so any potential gapping is out of the line of sight.

3) Avoid cuts that require a cope at either end, at least until you've honed your skills at making very accurate inside measurements.

To make the most effective and economical use of materials, rough cut (measured length plus 1½ inches) the longest runs first and set them aside. Use cut pieces of molding on shorter runs. Even the smallest piece may have its place. Set aside material with defects for use in closets or other areas where the defects won't be noticeable.

Figure 11-10 shows how I'd install the crown for a sample room. Keep referring back to the drawing as I take you through the sequence.

1) Piece 1 goes directly opposite the entry door. I simply butt to the walls at either end. It's not necessary to fit it precisely. First tack the ends of this and other runs to receive coped joints loosely. This allows them to conform more readily to the profile of the cope joint.

2) Cope piece 2 and piece 4 to the first piece. Note that any gaps in the coped profiles won't be visible from most angles in the room.

3) Back miter the free end of piece 2 so it becomes the underlying scarf joint. From the doorway, this joint will appear tight, even if a gap develops.

4) Cut piece 3 with the overlaying miter, and butt the other end to the wall.

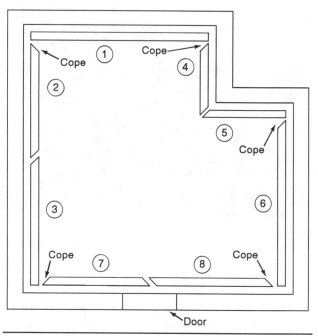

Figure 11-10   **Steps to installing a crown molding**

5) With the end of piece 4 coped and tested for fit against piece 1, mark the outside miter at the corner. Cut to the line and install the piece.

6) Miter and test-fit the end of piece 5 where it meets the miter of piece 4. Plane it to make a perfect fit and then install it. The opposite end butts the wall.

7) Cope piece 6 to fit piece 5. Butt the opposite end to the wall.

8) If the room isn't too wide, you can install the closing pieces (7 and 8) as a single piece. In this case, cut cope joints in either end. Make the piece slightly overlength and spring it into place. If you feel the room's too wide, make a scarf joint over the door opening. Be sure to cut and install piece 7 so it underlies the scarf miter first.

9) Cope the closing piece 8 to fit against piece 6. Then hold it in place and mark for the position of the scarf joint. Glue and nail through the joint itself.

That completes the crown molding. After final inspection, you're done.

Manhours for Cornice Moldings

All figures in the tables are in manhours and are based on the following assumptions:

- Tools and materials needed are available on site.
- The tradesman is a qualified and motivated finish carpenter.
- Work is good quality, stain grade, done no more than 9 feet above floor level.
- All defects are remedied before the carpenter leaves the site.

Add extra time for setup, cleanup, painting or staining, protecting adjacent surfaces, complicated layout or inadequate plans, repair and replacement jobs where fitting and matching is required, working around other trades, setting up scaffolding and ladders for work above 9 feet, and supervision, if necessary. Paint-grade work will usually reduce the time needed by from 20% to 33%.

Cornice or Crown Molding

Includes layout and coped joints, per linear foot

Up to 2¼" wide	.05
Over 2¼" wide to 3½" wide	.06
Over 3½" wide	.07

Chapter 12

Wall Treatments

In this chapter we'll talk about the most common wall treatments that finish carpenters install. These include tongue and groove wainscot, frame and panel wainscot, and sheet paneling. We'll also go over the cutting and fitting of wall trim such as chair and picture rail. I'll cover baseboard installation as a separate subject in the next chapter.

With the exception of ¼-inch sheet paneling, the wall treatments we'll be talking about here, including the trim elements, are installed after the standing moldings are in.

Remember, we're still using the "top down" rule. That means ceiling treatments are already in, except for cornices falling on full height wall paneling. Any protrusions, such as chimneys, heaters, utility vents and chaser boxes, or built in cabinetry, should also be in place.

Now's the time to remove any scaffolding that's still in the room. Sweep out the cutoffs. Then set up the chop saw and portable table saw in the middle of the room.

Tongue and Groove Wainscot

Tongue and groove wainscot (as in Figure 12-1) is generally made from fir or pine boards 3 to 5 inches wide and from ⅝ to ¾ inch thick. The traditional height for wainscot is the point where the cap rail would meet the back of a chair pushed against the wall. That's about 3 feet above the finished floor. Modern chair rail is usually installed one-third the height of the wall. In any case, follow the architect's instructions.

The first step is to mark the height of the wainscot on the wall with a perfectly level line. If you don't have a builder's level, use a reservoir-type water level. I'll talk about setting up and using this type of water level in a later chapter.

Cut the wainscot to a length that's ¼ inch less than the distance between the floor and the level line. This ¼ inch allows for variations in the floor. Don't worry about the gap at the bottom. It's sure to be covered by baseboard.

Nail the wainscot to the wall shoe at the floor line and along the top edge of the wainscot. If you're lucky, the framers will have installed blocking between the studs at the correct height. If not, you'll have to install a running backer as a nailer for the top of the molding. To install this backer, cut through the drywall. (Most codes require drywall under wood wainscot.) Then nail in the 2 x 4 blocking. Another alternative is to add a layer of ½-inch underlayment-grade plywood to the wall on top of the drywall.

Check the wall thickness carefully before you decide to add plywood backing. It could bring the wainscot out from the wall well past the standing moldings. This would ruin the finish appearance.

Be sure to ask the painter if he wants to finish the wainscot before it's installed. If not, suggest that he prime or seal the back of the stock to help stabilize the wood. In any case, always paint or stain the tongues before installation. Otherwise the raw edges will show up like neon lights as the wood shrinks and the tongues pull back out of the grooves.

Photo by Pat Cudahy

Figure 12-1   **Tongue and groove wainscot**

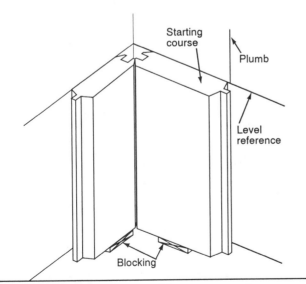

Figure 12-2   **Starting board in tongue and groove wainscot**

Installing Tongue and Groove Wainscot

As a rule, I start installing tongue and groove wainscot in a left corner and work my way toward any standing moldings. Since I'm right handed, this makes nailing easier. I can hold the board in place with my left hand and nail it with my right hand. Left-handers may find working right to left easier. In either case, avoid working toward any corner where you would have to fit the last board against an uneven wall surface, such as stonework. It's better to start at that corner, even if you have to work right to left. Concentrate on fitting the wainscot board edge precisely against the uneven wall. The other side of that piece of wainscot — the open side — just needs to be plumb.

If you have to miter any outside corners, preassemble each corner and install the molding as a unit. That way you can work *away* from this corner in either direction.

Now let's go through how to install tongue and groove wainscot step by step. We'll start at a corner at the left end of a wall and work toward the right. The left end of this wainscot butts up to another run of wainscot at the corner. Here we go:

1) Cut the stock to length and hold it to the wall, tongue side to the right. Hold the board away from the wall a little less than the thickness of the stock. That way, the starting course for the meeting wall won't look wider than the first starting board.

2) Be sure the top is even with the level layout line, and the outside edge is plumb. See Figure 12-2. Hold a level

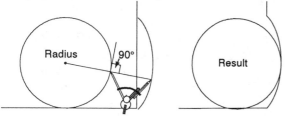

A Holding scribe parallel to radius of curvature (perpendicular to the surface)

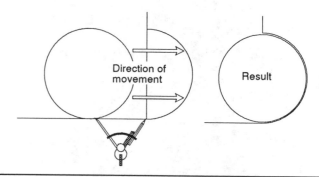

B Holding scribe parallel to direction of movement

Figure 12-3 **Scribing principles**

against the tongue on the right. When the piece is plumb, nail it to the wall's base shoe and the running backer.

3) Where there's no wainscot butting against the starting course, you'll probably have to trim the left edge of the starting course so it fits tight to the wall when the right edge is vertical.

Hold the piece plumb with the left edge about ½ inch from the wall.

Shim it up so the top edge is on the level line.

Set a pencil compass to span the maximum gap between the wall and the edge of the board.

Give it another ⅛ inch or so to keep the pencil from catching on the edge of the board.

4) If the wood is dark, you won't be able to see the scribed pencil line. To make the line more visible, lay down a strip of painter's masking tape along the edge to be marked. Then mark on the tape.

When you draw the scribe line down the wall, be sure to hold level the two ends of the compass. If one end or the other moves ahead too fast, your scribe line won't be parallel to the wall.

Figure 12-3 shows what happens if you don't hold the two ends level. In Figure 12-3A, the two ends of the compass don't stay level as the arc is drawn. Instead, the ends are on the extended radius line. That's the natural tendency when drawing an arc around some irregular object. The result is an arc with a shape that doesn't match the original, as you can see.

Figure 12-3B shows the correct procedure. Notice that the arc drawn matches the original perfectly. Keeping the ends of the compass level is the trick. So curb that tendency to hold the compass perpendicular to the shape being traced.

5) Cut the first board to shape and nail it in place. Then fasten the rest of the wainscot run to the wall. Nail through the tongues into the backing boards. If you use a hammer to seat the groove over the tongues, use a scrap block of wainscot to protect the tongue. Check every fourth or fifth board for plumb and adjust it before nailing.

6) You'll likely have to rip the closing board to fill the gap between the second-to-last board and the wall corner. If the meeting wall is distinctly bowed or irregular, use a pencil compass to scribe the cut line.

Measure the gap between the top of the last board and the wall, setting the compass to this width.

Photo by Pat Cudahy

Figure 12-4 **Scribing the closing tongue and groove board**

Measuring from the top left edge, transfer your compass measurements to the face of the closing board and make a mark.

Do the above procedure at the bottom of the board, making a measurement mark with the compass.

7) Now hold the marked board in the corner. Hold the compass level and set it so the needle end touches the wall while the pencil touches the top mark. Keep this setting and move the compass to the bottom of the board. Adjust the bottom of the board in or out until the pencil touches the lower mark. Now drag the pencil compass from one mark on the closing board to the other. Make sure you hold the legs of the pencil compass level as you do this. See Figure 12-4.

If the wall is out of plumb, but straight, you won't have to use the scribe. Just connect the top and bottom marks with a straightedge.

8) Now cut to the line with either a small-bladed circular saw or a jig saw. You'll need a back bevel of 2 to 3 degrees. If necessary, review Figure 10-6 in Chapter 10 on cutting a back bevel. This bevel lets the closing board swing into place without restriction. And the sharp edge of the back bevel makes it easier to plane off any excess when you make the final fitting.

Frame and Panel Wainscot

Occasionally you'll have to plan the frame and panel layout for wainscot. You may have to decide on the location of each frame and panel. This depends on the length of the wall runs and position of windows and other structural members where the paneling will be installed.

With any luck, the layout will be specified in the architect's elevation view of the room interior. If not, draw your own scaled sketch of the panel layout. That way you can see if the proportions of panel width to panel height look right.

How high and wide should the panels be? The ancient Greeks believed that the most pleasing proportions for any rectangle were a width-to-length ratio of 5 to 8. This was called a "Golden Rectangle," and much of Greek architecture was based on this principle. Of course, every panel doesn't have to be ideal. Windows, and the need to evenly space panels along a wall, make compromise necessary.

Figure 12-5   **Frame and panel wainscot layout**

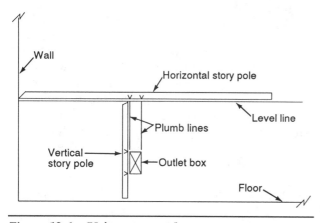

Figure 12-6   **Using story poles**

Another rule on proportion is to keep panels as simple, four-sided rectangles rather than shaping them around protrusions. Windows, for example, should have complete panels grouped beneath them, as shown in Figure 12-5.

Proportion the frame around panels so adding the baseboard won't make the bottom rail look narrower than the top rail. Note in Figure 12-5 that the bottom rail is considerably wider

than the top rail. When you install the baseboard, the exposed rail will come out the same width as the top rail. Also, make the stiles at the ends of a run at least the same width as the stiles separating the panels.

If you plan to fabricate panels away from the site, you'll need story poles. They give you a full scale reference to the layout of the width and height of the panels and the overall size of the assembly (Figure 12-6). You can mark the horizontal and vertical poles to indicate positions of standing moldings, outlet boxes, heating ducts, and any other objects that you have cut around for the panel assembly.

Making Frame and Panel Wainscoting

Frame and panel wainscot is usually a solid wood framework joined around either a raised or a recessed panel. If you're going to paint the wainscot, use panels made from MDF, which is a high-density particleboard. Otherwise, use a raised panel made from solid wood, and a recessed panel made from a ¼-inch veneered panel.

The most common way to join the rails and stiles of the frame is with a shaped stick and cope joint. This joint, which you often see in cabinet door frames, is created with specialized cutting bits on a shaper. If you don't have access to a shaper, you can make joints with spline biscuits or dowels.

If someone else is fabricating panels off site, make these suggestions to the fabricator:

1) The grain of the panels, especially if stain grade, must run in a uniform direction, usually vertical. This is because a room with wainscot will generally have paneled partition doors, and the grain of these doors always runs vertical.

2) Preassemble wainscot frames to the full length of the run. That way the installer won't have to deal with drawing together joints in the middle of a wall.

3) Leave end stiles meeting the wall, and stiles abutting standing moldings, oversized so the installer can cut and fit them. If a run fits between two walls and the ends aren't covered by another run of wainscot, the fit to the walls has got to be perfect. In this case, omit a stile at one end of the run. Right handers will vote for omitting the right end. This end can then be fitted and installed on site.

4) Layout of utility boxes, heating ducts, and other openings in the panel are probably on the story pole. Even so, don't let the person making your wainscot cut these openings. There's too much room for error. Cut these openings on site.

Installing Frame and Panel Wainscot

As with tongue and groove wainscot, begin by making a level line on the wall to mark the top of the paneling. Start at the wall at the left side by setting a panel in place. Use blocking to hold the panel level. Now, scribe and cut the panel to fit exactly in the left corner. This is the edge you'll run the tape measure or story pole from to determine other cuts. It's important to try for a perfect fit here. Test the fit, then lift the panel away from the wall and lay it face down on a pair of sawhorses. With the panel on the sawhorses, make your cuts for outlet boxes and other openings.

Determining and cutting out the outlet boxes-
As I mentioned earlier, I determine the location of outlets and other cuts by making up my own pair of story poles. I find using marking sticks (story poles) is more foolproof than measuring with a tape. Be careful here! A miscut on an 8-foot length of frame and raised panel wainscot can cost a lot of time and money.

Cut ¾ x ¾ inch story poles from a straight length of stock. Cut out two poles, one for the horizontal marks, and one for the vertical. Make each pole a little longer than the longest measurement you'll have to make.

Figure 12-6 shows how to locate and cut out for an outlet box:

1) Use a level to plumb lines up from the outlet box sides to the level line drawn on the wall.

2) Run one stick horizontally along the level line toward the corner of the opposite wall. Be sure the point of the stick is tight to the wall.

3) Mark the stick where the plumb lines intersect the level line.

4) Lift the stick away from the wall and lay it on the back of the wainscot along the top edge. Use spring clamps to keep the stick tight to the wainscot and flush to the edge that butts the corner. Transfer the marks to the wainscot.

5) Use a square to draw the lines out from the marks across the back of the panel.

6) Hold the second stick against the wall and parallel with one of the plumb lines running to the outlet box side. Be sure

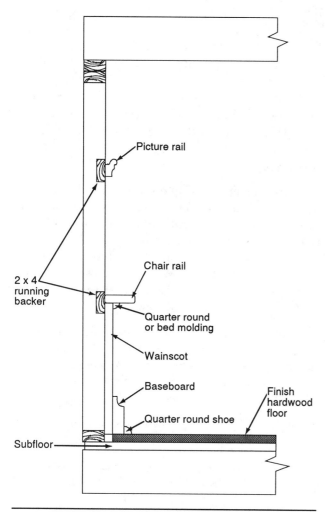

Picture rail

Chair rail

2 x 4
running
backer

Quarter round
or bed molding

Wainscot

Baseboard

Finish
hardwood
floor

Quarter round shoe

Subfloor

Figure 12-7 **Wall cross-section of picture and chair rail installation**

the top of the stick is even with the level line. Mark the top and bottom of the outlet box on the stick. Unless the box is severely out of level, you'll only have to measure along one side.

7) Carry this stick to the wainscot. Hold the top of the stick flush to the top edge of the wainscot and parallel with one of the square lines you just drew. Transfer the outlet box top and bottom marks to the wainscot.

8) Square across the marks to the other line. Then hold your breath and cross your fingers! Drill a hole the diameter

of a jig saw blade in one or two of the corners of the rectangle to give the saw a starting place. Cut out the box hole.

9) Attach the wainscot to the wall.

Use either finish nails for paint-grade work or screws for stain-grade for attaching wainscot. If you use screws, cover the screw heads with wood plugs. Cut these plugs from the same material as the stiles. Be sure to set the plugs with the grain parallel to the grain of the rail or stile.

Now, if you've been paying proper attention here, you know we've still got to fit the right side stile of this wainscot unit to the wall. You'll remember that we had this stile left off because the run has to fit between two walls. Now's the time to cut it to fit. Use the same procedure I explained earlier in this chapter for fitting the closing board of tongue and groove wainscot.

Installing Chair and Picture Rails

Figure 12-7 shows the cross section of a wall featuring a typical wainscot with a chair rail and picture rail. Note the running backings — in this case 2 x 4s set between the studs — installed behind the rails. This makes it easy to locate your fasteners. It also lets you draw the molding tight to the wall at any point. And believe me, if the drywall has noticeable humps or valleys, that backing can be a godsend!

After you install the wainscot, attach the chair rail to the top edge of the boards or the top rail of the frame and panel assembly.

Sometimes you'll be installing chair rail without wainscot. In that case, install the rail to a level line drawn on the wall. Picture rail is, of course, always installed to a level reference line.

To install both chair and picture rails, start in a corner and work out toward any standing moldings or to the other end of the wall. Miter

Figure 12-8   **Notched chair rail**

outside corners and cope inside corners. If the chair rail is simply a flat piece of stock with a molded edge, you can get away with mitering the inside corners as well.

Since chair rails often protrude further out from the wall than the window and door casings, you'll have to notch them around such moldings. See Figure 12-8. Return miter a fully exposed end — one not butting a standing molding or room corner.

Chair rails are often made from more than one molding. The cap may be supported from below with a bed molding (as in Figure 12-7), or by small cornice molding. The top may be molded with another bed mold or a quarter round. Install these moldings after the cap is in. Cut a miter on outside corners and cope the inside corners.

Install the chair rail and any extra moldings with finish nails. But don't rely on finish nails to hold a picture rail. That's especially true if there's any chance the rail will actually be used for its intended purpose — supporting pictures! To be safe, attach the picture rail through to the studs with 2-inch drywall screws. Countersink the screws and cover the screw heads with bungs as described in Chapter 10.

Installing Sheet Paneling

Most sheet paneling is ¼-inch-thick veneer plywood panels installed from floor to ceiling along the length of a wall. Unless the walls are concrete or masonry, the sheets are applied over a layer of drywall. To do a first class job, the sheets must be fitted carefully to one another and the wall surfaces they abut.

The first step is to get the surface ready for installation.

Preparing Walls for Sheet Paneling

Make sure the running backings are in place and properly located. As I mentioned in Chapter 6, running backings are generally 2 x 4s set between the studs. Use at least two, and make the first one at hip height. That gives the panels a feeling of rigidity. Space the second evenly between hip height and the ceiling.

Now check to see if the studs fall on 4-foot centers. If they don't, you have to install backing for joints on the 4-foot-wide panels. One solution is to fur out the wall so you're certain there are nailing points at all joints.

For furring applied to a concrete or masonry wall, nail furring strips to the wall with a powder-actuated masonry nail gun. Be sure to cover the wall first with mastic and then apply a plastic vapor barrier. This keeps water vapor away from the back of the paneling.

Use only dry wood for the furring. Run it horizontally on 16-inch centers. Install the vertical strips on 4-foot centers to be sure there's support behind each panel joint.

The furring should correct for any humps and hollows in the wall. Place shims under the furring at the low points you've discovered by holding a long straightedge against the wall. After you install the furring, check both vertical and horizontal members with the long straightedge, and adjust the shims if necessary. Figure 12-9 shows the completed job.

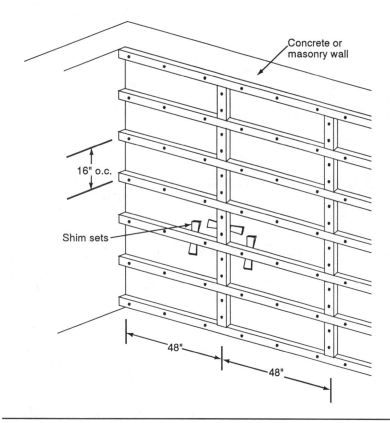

Figure 12-9 **Paneling furring strips**

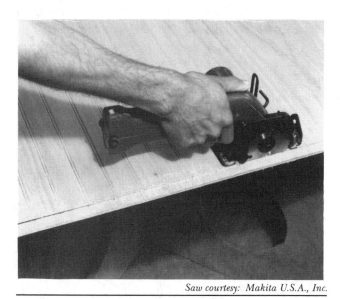

Saw courtesy: Makita U.S.A., Inc.

Figure 12-10 **Using a cordless panel saw to cut paneling**

You can also apply paneling directly over foam insulation board. In this case, you don't need any furring. Use a construction grade panel adhesive and glue the paneling directly to the insulation board.

Fitting the Panels to the Wall

Begin fitting by cutting the paneling to height. I usually cut it ½ inch shorter than the distance between the finish floor and the ceiling. That allows a little room for adjustment. The gap will be covered by moldings.

You can cut three or four of the sheets at once by clamping them to a sheet of plywood supported on a pair of sawhorses. Face the good side down. That way, the back of the paneling, not the front, gets any tearout from the saw. Facing the good side down also keeps it from being scratched under the saw. On single sheets of paneling, a cordless panel saw is a lot easier to work with than a standard circular saw. See Figure 12-10.

Figure 12-11 shows paneling that ends at an irregular surface, stone veneer in this case. A full 48-inch-wide panel would just touch the veneer in some places. But it wouldn't fill the point of maximum indentation without moving the other side off the vertical nailer. So instead, we'll fill in with a panel cut to fit into the point of maximum indentation and land on center of the next nearest vertical nailer. Measuring from the center of the stud, we find this is 34½ inches. See Figure 12-11.

Here's how to install the panel that ends at the stone veneer. Follow along in Figure 12-11 as I explain the procedure.

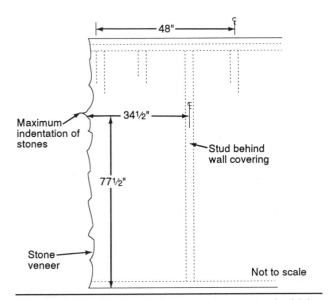

Figure 12-11   **Determining maximum panel width**

1) We'll begin by cutting the panel to rough width. Measuring from the center of the stud to the furthest point the panel will have to reach, we find that the widest point is 34½ inches. So cut the sheet to a width of 34¾ inches. The extra ¼ inch provides a surface for the compass pencil to ride on without catching on the edge.

2) Next, we want to mark the point of maximum indent on the panel. That point is 77½ inches above the floor. But the panel will be raised ¼ inch off the floor. So mark the panel at 77¼ inches above the bottom of the panel and 34½ inches from the right edge.

3) Move the panel right up against the veneer and support it ¼ inch above the floor on scraps of ¼-inch sheeting. Adjust your shims so the right edge of the panel is plumb. Now tack the panel temporarily in place.

4) Set your pencil compass to the distance between the maximum indent and the mark you just made on the panel. Run your compass down the full height of the stone, marking a line on the panel that duplicates the edge of the veneer.

As I've warned already, make sure you hold the compass legs level when marking this line.

5) If the panel is dark, apply a strip of wide masking tape near the edge and mark on this tape. Don't leave the tape on any longer than necessary. After about a half an hour or so, it's difficult to remove it cleanly.

6) Remove the panel from the wall and use story poles to mark utility boxes and other cutouts. Make the marks on the back of the panel. Another way to find the location of utility boxes is to chalk the edges of the boxes. Then hold the panel in place and press the panel against the chalk. Enough chalk will transfer to the back of the panel to show the cut line.

7) Cut along the scribed line with a jig saw and test the fit. If it looks good, and there are no other cuts to be made in the panel, it's ready to install.

Installing the Paneling

Install panels to the walls with threaded, painted head nails. Space them vertically every 16 inches over studs or furring strips. At the joint with another panel, reduce the spacing to every 8 inches.

Make sure every panel fits tight before starting on the next panel. Sometimes a little extra pressure is all that's needed to get a tight fit. Other times you'll have to plane the edge of the panel to get a good match at the joint. As a precaution, I often paint the area behind the joint with flat black paint. If the joint opens up over time, or if I can't get a snug fit the entire length of the joint, the black background will hide my offense.

Once the paneling is up, install any cornice and base molding that's needed. Then apply the standing moldings around the door and window jambs.

I'll cover installing base moldings in the next chapter.

Manhours for Wall Treatments

All figures in the tables are in manhours and are based on the following assumptions:

- Tools and materials needed are available on site.
- The tradesman is a qualified and motivated finish carpenter.
- Work is good quality, stain grade, done no more than 9 feet above floor level.
- All defects are remedied before the carpenter leaves the site.

Add extra time for setup, cleanup, painting or staining, protecting adjacent surfaces, complicated layout or inadequate plans, repair and replacement jobs where fitting and matching is required, working around other trades, setting up scaffolding and ladders for work above 9 feet, and supervision, if necessary. Paint-grade work will usually reduce the time needed by from 20% to 33%.

Wall Treatments

Manhour estimates for paneling do not include molding installation. See manhours for running moldings in Chapter 13.	
Board wainscot or paneling, 3½" to 5½" wide tongue and groove, per square foot of wall covered	.08
Frame and panel wainscot, installation only. Add for fabricating the frames and panels. Add for running moldings, if required. Manhours per square foot covered	.09
Add for frame attached with screws and covered with wood plugs	.01
Sheet paneling installed on wood-frame or masonry walls. Add for running molding, if required.	
Manhours per square foot	.04
Add where vapor barrier must be installed with mastic over a masonry wall	.01
Add where 1 x 4 furring is nailed on masonry wall with powder-actuated nail gun. No shimming or straightening included. Manhours per square foot of wall	.02
Add if sheet paneling is scribed around stone veneer, per linear foot scribed	.15
Minimum time for scribing sheet paneling	.50

Chapter 13

Base Molding

Many homes built today have no window casings, no cornices, and no door moldings to speak of. There are no stair balustrades or wooden stair treads, and there's no woodwork on the walls. If every home didn't need baseboard to protect vulnerable wall surfaces from damage, we finish carpenters might go extinct! That's reason enough for you to get good at running this stuff.

Preparing to Run Baseboard

When you're working "top down," baseboard goes in last. By the time a house is ready for it, the ceiling and wall treatments, standing moldings, built-in cabinetry, fireplace surrounds, and nearly every other task on a finish carpenter's job sheet is already done. The hardwood flooring is in. If not, the flooring people will probably ask that you leave the baseboard off until they're done installing, sanding, and finishing.

You can nail narrow baseboard into the bottom wall plate, but you'll have to nail wider baseboard higher up on the wall. If you were on site when the walls were framed, you could have added backer strips on the wall where needed. Otherwise, you'll have to determine where the studs are so you can nail the top edge of the baseboard into them. To make it easy, I mark the location of wall studs on the floor. Even if I feel enough running backing is in place, I still take the time to chalk the location of the studs on the floor around the entire perimeter of the room.

To get ready for installation, set up your chop saw station in the center of the room. Keep baseboard stock at hand but out of the way. With your kneepads on and your running shoes snugly tied, you're ready to go.

Fitting Baseboard to the Walls

Making joints in baseboard is similar to making joints in crown molding. You cope the inside corners, miter the outside corners, and extend the lengths with scarf joints. (You might want to go back and review Chapter 11 on those points.)

To make things even easier, the sequence for installing baseboard follows the same rules as those for crown moldings. Figure 13-1 is, in fact, the same room you saw in Chapter 11. The only difference in layout, sequencing, and installing the base molding is the break around the entry door.

Miter Joints in Baseboard

While most of what you learned about crown moldings is true of baseboard, there are some differences in fitting baseboard joints.

Often, the outside corner of a room won't be plumb and it won't form a perfect 90 degree angle. Since the baseboard is designed to sit flat against the wall, this is a problem when fitting corner joints.

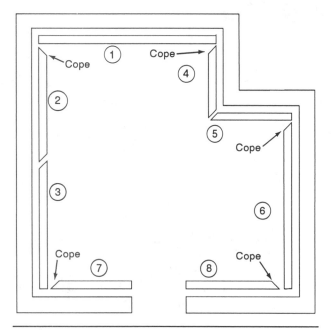

Figure 13-1   **Baseboard layout**

Cornice moldings, on the other hand, sit at an angle to the wall corner. So they're relatively unaffected by an out-of-plumb corner line.

To find the bisecting angle when the wall doesn't meet at a 90-degree angle, lay a sample of the baseboard against each face of the wall. Draw a line at each outside edge, as shown in Figure 13-2. Next, draw a line joining the corner of the wall to the intersection of the two lines. That's your bisecting angle.

Transfer the angle to the baseboard by setting the molding in place and squaring a line up from the intersection point. Draw a line from the corner of the wall to the square line where it appears at the top of the baseboard. See Figure 13-3. This is harder if the top of the molding is profiled. In this case, use a protractor to read the angle directly from the lines on the floor. Set the chop saw to this angle when cutting the baseboard to length.

With a noticeably out-of-plumb corner line, you'll have to make a choice. Your first option is to cut the miter at a compound angle (the miter angle plus the angle the corner leans away from plumb). Your other choice would be to cut the miter square, and let it stand away from the corner.

If the corner leans out, there won't be a problem with cutting the miter square. The molding's top edge then lays tight to the wall. But if it leans the other way, there'll be a gap along the top edge of the baseboard. In paint-grade work, you can fill the gap with caulk or Spackle. But with stain grade baseboard, especially where the base won't have an additional cap piece, you've got to cut the miter at a compound angle. To find the wall angle portion of the compound miter, place the body of the bevel gauge along the top edge of the molding running past the corner. Set the blade to the wall.

Figure 13-2   **Outside corner laid out for bisecting angle**

Figure 13-3   **Square line drawn on face of baseboard from intersecting point**

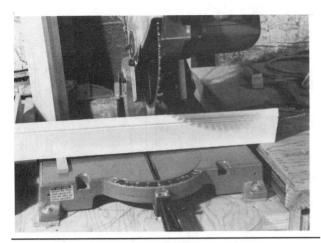

Figure 13-4 **Setting a standard chop saw to cut compound miters**

You don't necessarily need a sliding compound miter saw to cut compound angles for this job. A standard chop saw will do as well. Lift the baseboard on a scrap of wood to one side of the blade. See Figure 13-4. Slide the scrap toward, or away from, the blade to fine-tune the angle. Hold the base firmly to the back fence and then make the cut. Practice cutting the angle on some scraps before cutting the baseboard itself.

Butt Joints in Baseboard

Because of standing moldings, there are a lot more visible butt joints in baseboard than there usually are in cornice moldings. To determine the location of a butt joint, you can run the stock by standing molding (the opposite end of the baseboard should already be fitted) and make a mark at the top edge of the baseboard. Cut the molding slightly long to the waste side of the line and check the fit. This method works well enough if the standing molding is plumb, or if the baseboard is narrow enough not to show much of a gap if the angle is off.

On the other hand, if you're running wide baseboard, I recommend using a *preacher* to get the cut just right. He knows just what prayers to offer. Just kidding. A preacher is a shop-made marking tool that automatically transfers the position and angle of standing molding to the baseboard. See Figure 13-5.

Make the preacher from hard, split-resistant wood or a scrap of hardwood plywood. Notch the wood just slightly wider and higher than the molding stock. For the tool to work properly, the edges must make a perfect 90-degree angle with the face of the hook.

Slip the preacher over the baseboard and hold it firmly to the edge of the standing molding you wish to cut the baseboard to. Make sure the preacher is square to the wall, then draw a line onto the face of the baseboard along the edge of the tool. This line is a perfect reflection of the standing molding.

Photo by Pat Cudahy

Figure 13-5 **Using a preacher tool**

Dealing With Open Ends

Baseboard that ends in the middle of a wall or butts against a smaller molding must somehow end neatly. There are a couple of good ways to do it.

1) Return-miter the end into the wall. To do this, use exactly the same method as I showed earlier for return-mitering the ends of the aprons running under window stools. See Chapter 9, Figure 9-15.

2) The second method is to return-miter the baseboard into the floor. Figure 13-6 shows an example. This trick makes the most of an awkward situation. Thanks to finish carpenter Dennis Calleson for this detail!

Installing the Baseboard

Attach the baseboard to the walls with a pair of finish nails driven into each stud. If you're lucky enough to have backing, you generally won't need to find the studs. Simply nail the molding off every 16 inches. The lower nail goes into the wall shoe, the upper nail goes into the backing.

Figure 13-6   **Baseboard return-mitered to floor**

If you're installing the baseboard over an uneven floor, common with masonry or tile, you may need to scribe the bottom edge in some places to meet the surface without unsightly gaps. Here's how to scribe a length of baseboard.

1) Find the wall where the gap between the leveled baseboard and the floor is the greatest. When the baseboard is fitted, the top of the molding at this point becomes the standard you scribe all other runs to.

2) Set the baseboard in place. Use small wedges to lift the baseboard slightly off the floor.

3) Check to see that the top is level. Adjust your wedges until it is. Then lightly tack it in place at either end.

4) Now set your pencil compass to the depth of the gap at the deepest point.

5) Use your pencil compass to draw the scribe line down the entire length of baseboard. Be sure to keep the legs of the compass perfectly vertical.

6) Now cut the baseboard along the line you drew.

7) Install this baseboard and repeat the process for the other runs. Be sure to set the scribe so the top of the other boards, when fitted to the floor, come even at the top edges.

Installing Baseboards Over a Solid Floor

When you install baseboard in a room with a wood floor, insert $3/16$-inch-thick spacers between the molding and the floor. See Figure 13-7. After you nail the base off to the wall, be sure to remove the spacers. If you don't, any swelling or cupping of the floor is likely to move the baseboard as well.

To cover the gap between the floor and the baseboard, install base shoe between the baseboard and the flooring. Try to nail the shoe only into the baseboard, and not to the flooring.

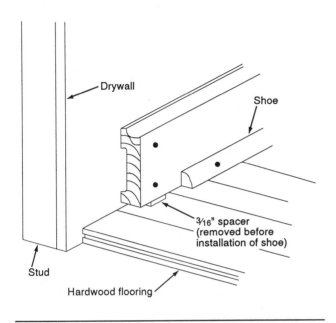

Figure 13-7 **Baseboard cross section**

Labels: Drywall, Shoe, Stud, Hardwood flooring, 3/16" spacer (removed before installation of shoe)

If the shoe is attached to the floor, a gap may open between the shoe and the baseboard if the flooring shrinks, pulling back from the wall.

Installing Baseboards to Curved Walls and Rounded Corners

Life doesn't always go in a nice straight line, and neither do the walls of many a house. When either life or an architect throws you a curve, it always pays to have some tricks up your sleeve.

In life, we learn to cope with these curves. With baseboard, we have to kerf — that is, cut kerfs to get the molding to go around a curve. Kerfs are simply saw cuts run into the back of a molding to within ⅛-inch of the face. You can do this easily on a sliding compound miter saw by setting the depth of cut. See Figure 13-8. You can also use a circular saw.

Spacing the kerfs depends on the radius of the curve. Figure 13-9 shows a graph for spacing kerf cuts. These figures assume cuts are made ⅝-inch deep in a piece of ¾-inch oak. Figure 13-9 shows spacing of kerfs for each curve radius.

These kerfs make it easy to bend the baseboard into what looks like an even curve, while not risking breaking the board.

When in doubt about how much you can bend a wood species, make a few test kerfs in a piece of scrap.

If you have any doubts, space the kerfs closer. This reduces the width of the flat, uncurved, sections. If you're going around a tight corner, wider flats become quite noticeable. When you nail kerfed molding to the wall, drive nails in the center of flat sections, *not* in the kerfs.

Kerfs work best in plain rectangular molding. You can cut kerfs into most profiled moldings if you stop the kerf cut before it breaks through the face along the top profiled edge. But a better choice is to kerf and bend a plain base molding and then cap it with a small profiled molding that's easy to bend without kerfs. Most cap molding will bend around curves with a radius greater than 3 feet without any help.

Here's another trick to keep up your sleeve. If you need to run the cap molding around a tighter bend, steam the molding in a length of metal stove pipe. Cap one end of the pipe. Prop up the pipe at a 45-degree angle with the capped end down. Slide the molding into the top end of the pipe and plug it with a rag. Open the cap at the bottom a little and feed steam from a boiling kettle into the lower end of the pipe. This turns the pipe into a steam pressure cooker.

Figure 13-8 **Making kerf cuts using a compound miter saw**

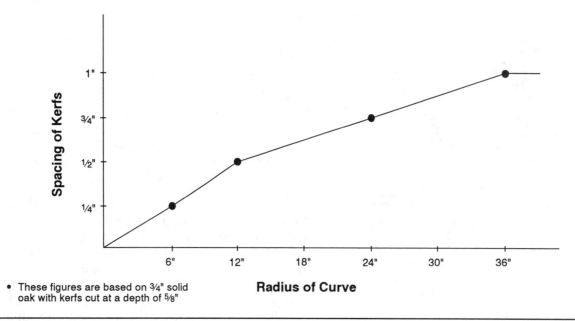

- These figures are based on ¾" solid
  oak with kerfs cut at a depth of ⅝"

Radius of Curve

Figure 13-9　**Kerf intervals**

Let ¾ x ¾ inch molding sit for about 20 minutes. Then test it on the curve. If there's a lot of resistance, steam it for another 10 minutes with more boiling water. The amount of steaming depends on the kind and size of wood you're using, the size of the molding, the moisture content and the temperature of the steam. An important word of caution: Don't steam oak in steel pipe. It turns the wood black! Instead, use a length of PVC plastic drain pipe. Don't worry about wetting the wood too much. The steam actually tends to dry the wood out!

Also, don't burn the building down. Stay close to the operation. Have a fire extinguisher handy just in case some bozo knocks over the rig.

There's another way to install molding around a curve without steaming and without kerfing. Use plastic. Obviously this won't work if the job calls for stain grade wood. But if you're installing paint grade molding, plastic moldings bend around fairly tight curves without distorting the profiles. There's a wide variety of plastic profiles available.

Dealing With Rounded Wall Corners

Rounded drywall corners are becoming more and more popular in modern homes. Fortunately for us finish carpenters, corner blocks designed to fit against standard corner rounds are readily available. See Figure 13-10.

Figure 13-10　**View of corner block**

Unfortunately, these corner blocks are usually only available in paint grade.

22½ degree mitered corners- You can make a perfectly acceptable stain-grade corner by trimming out the rounded corner with the baseboard itself. Cut a 22½ degree miter in the end of the baseboards coming to the corner. Then fill in the space with a short length of baseboard. See Figure 13-11.

This isn't a true rounded corner. But it's a reasonable substitute and much cheaper than having rounded corner blocks specially produced in solid wood.

Figure 13-11 **22½ degree turned corner**

Manhours for Base Molding

All figures in the tables are in manhours and are based on the following assumptions:

- Tools and materials needed are available on site.
- The tradesman is a qualified and motivated finish carpenter.
- Work is good quality, stain grade, done no more than 9 feet above floor level.
- All defects are remedied before the carpenter leaves the site.

Add extra time for setup, cleanup, painting or staining, protecting adjacent surfaces, complicated layout or inadequate plans, repair and replacement jobs where fitting and matching is required, working around other trades, setting up scaffolding and ladders for work above 9 feet, and supervision, if necessary. Paint-grade work will usually reduce the time needed by from 20% to 33%.

Running Moldings

Single piece mitered installations nailed by hand. Add 100% for two-piece molding. These estimates are based on stain grade work. Subtract 33% for paint grade work. Manhours per linear foot of molding

Picture rail	.05
Chair rail	.04
Wainscot cap	.04
Bed molding	.04
Baseboard	.04
Kneewall cap up to 8" wide and 1½" thick. Add extra time for layout and bed molding	.16

Chapter 14

Closet Interiors

Nearly all full-height closets will require at least a hat shelf and a clothing pole. Homes with large master suites, for instance, will probably require lots more. The plans will probably show some combination of site-built storage units or prefabricated multi-function fixtures with supports made of metal webbing. These metal units are generally installed by the suppliers. In this chapter I'll only deal with those closet units a finish carpenter may be expected to build and install.

Clothing Poles and Hat Shelves

A closet used primarily for hanging clothing should be at least 24 inches deep and high enough to hang the pole about 5 to 6 feet above the floor. Use a pole at least $1^5/_{16}$ inch thick and support it at every other stud — assuming 16-inch centers. Unsupported spans shouldn't exceed 36 inches.

The cleats and fixtures supporting the pole also carry the hat shelf. Make this shelf at least 11 inches deep from ¾-inch sheet or solid wood stock. Don't use pine boards sold as "shelf stock." The knots or sap pockets can bleed through the finish and stain fabrics or other materials stored on the shelf.

I recommend using pre-edged and surfaced sheet stock made from high density particleboard or Melamine. It comes already ripped to shelf width. Lengths on these from your local supplier may be limited to 8 feet.

Installation

Begin installation by marking level lines around the perimeter of the closet at the height of the underside of the shelf. Mark the location of stud centerlines just above this line.

Make up cleats from softwood 1 x 4s. Fasten these just under the level line by driving a nail or screw at each stud. Use either 12d casing nails or 2-inch trim screws. To save time, cut out all the cleats for all the closets at one time.

Now install metal support brackets to hold the shelf and pole. Some brackets are designed to install around the cleat. For a better-looking and stronger support, use a bracket with an integral, fixed back brace. With this type bracket, add a small vertical backing of 1 x 4. See Figure 14-1.

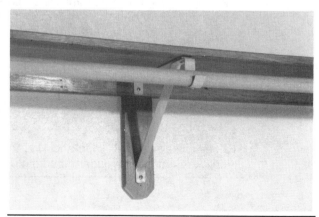

Figure 14-1   **1 x 4 vertical backing behind bracket**

If you use braced brackets, you can get away with not putting in the cleat running along the back of the shelf. Over time, though, the shelf is going to sag without that back cleat. That's especially true if the shelf is heavily loaded.

The design of the metal bracket determines where you place the pole support rosettes on the side cleats. Here's how to determine the correct location:

1) Hold the bracket against the side cleat and mark the position of the pole support on the side cleat.

2) If the closet is small and doesn't have intermediate brackets for reference, center the rosettes at least 13 inches away from the closet's back wall and 1¾ inches down from the top of the cleat.

Then cut the pole ³⁄₁₆ inch short so it slides easily between the pair of rosettes.

Cut the shelf to fit snugly between the closet side walls, unless it's a second shelf above the one that's at eye level. In that case, the fit won't be visible from below. You can cut the shelf a little shorter.

In most cases, closet corners aren't square and opposite walls aren't parallel. To make a good fit for the shelf, scribe the end cuts. To determine the cuts, hold a length of shelf stock cut slightly overlong snug to the back wall. Tilt one end of the shelf up while the other end rests flush on the 1 x 4 support cleat. See Figure 14-2A.

Most likely, only one edge of this shelf will be touching the closet side wall. Use your pencil compass to scribe the lower end of the shelf, the end resting on the side cleat. Remove the shelf and cut on the scribe line. Test it again. It should rest flush on the cleat and snug against the wall for the full length of the shelf.

A Rest one end of shelf on support cleat and scribe to fit.

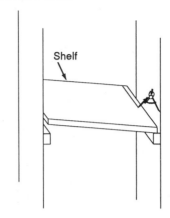

B Make length mark on shelf stock along back of shelf.

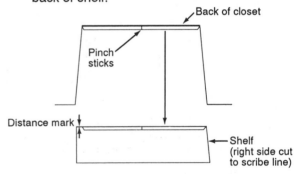

C Scribe a cut line on opposite end.

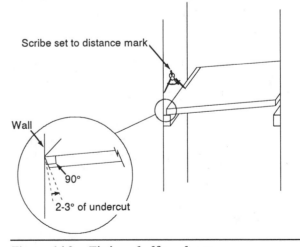

Figure 14-2 **Fitting shelf to closet**

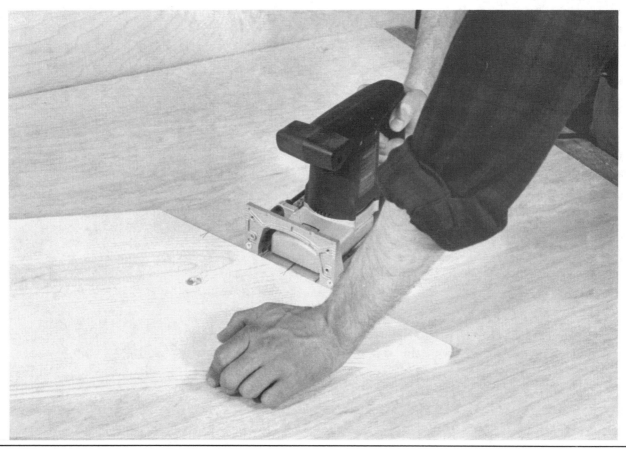

Figure 14-3   **Joiner slotting for biscuits along miter joint**

Use a set of pinch sticks to find the length of the shelf along the back wall. Make a length mark on the shelf stock at the back edge, as shown in Figure 14-2B. Return the shelf to the closet and then tilt the board onto the opposite side cleat. Set the gap on your pencil compass to the distance between the length mark and the wall at that point. Run your compass along the wall, as shown in Figure 14-2C. Cut to this line. The result should be a nice snug fit.

I like to add a few degrees of undercut to that last cut. This undercut makes installation easier and helps to keep the wall from getting scratched when you lower the shelf into place. See Figure 14-2C.

Install the shelf by tilting it onto the cleat along the wall first scribed. Then lower it onto the second cleat. Nail or screw the shelf to the support cleats. Holding them plumb, secure the metal brackets to the wall studs. Finally, screw the brackets to the shelf from below.

Shelving Around Corners

If the closet is L-shaped, the hat shelf and the pole need to go around a corner. I find it's easiest to join the shelving with plate biscuit joints. Figure 14-3 shows a joiner slotting for biscuits along the edge of the miter joint.

You can also join the shelving with dowels, a full-length spline, or a butt joint glued and screwed to a bottom cleat spanning the joint, as shown in the drawing in Figure 14-4.

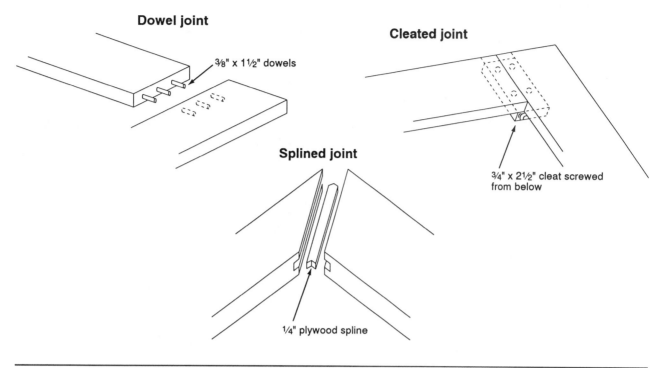

Figure 14-4   **Alternative shelf corner joinery**

Corner joints for closet poles- If the shelf and pole go around an outside corner, join the poles with a miter joint. Marking for miter cuts on round objects is never easy. My advice is to make your miters before you cut the poles to length. If you make a mistake, there's plenty of stock left to try again.

Figure 14-5   **Inside corner of poles joined with a cope joint**

It's best to join inside corners with a cope joint. See Figure 14-5. For this joint, I prefer to make up a jig to guide the drill bit. See Figure 14-6. This jig makes it easy to cut a half circle into the pole that's the same diameter as the pole stock.

To use this jig, slide one end of an overlength piece of pole into the jig's horizontal side hole. Clamp or hold it firmly in place. Insert your drill bit into the vertical hole and run it through the stock. See Figure 14-7. This provides a perfect cope joint.

Use pinch sticks to find the correct length for the pole. Then cut the pole to length. Glue and then screw the joint together from the back.

The shelf and pole are supported near the corner juncture with the usual cleat along the back edge of the shelf, and a pair of metal brackets. At inside corners, place the brackets at the first available studs anywhere from 12 to 24 inches away from the corner. At outside corners, place the brackets within a few inches of the corner juncture.

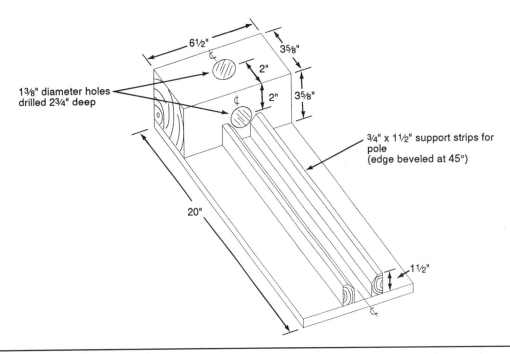

Figure 14-6 **Jig for coping closet poles**

Site-Built Closet Storage Unit

Figure 14-8 shows a typical site-built clothing storage unit for a master bedroom closet. The versatility and speed of the hand-held plate biscuit joiners make it easy to do this type of work on site. Slot joinery is a lot easier than trying to do precision dado or rabbeting work with table saws or radial arm saws set up on a construction site.

Doing work like this on site saves time and may be cheaper than farming the task out to a cabinet shop. In my experience, this kind of permanently-fixed case work with plate biscuit joinery is strong and durable.

Construction notes- You can construct the entire storage unit, including drawer boxes and base frame, from ¾-inch softwood boards, ¾-inch hardwood plywood, or ¾-inch surfaced Melamine shelf stock. Again, don't use pine because the sap tends to bleed, discoloring the paint and even ruining clothes. The only hardware you need are adjustable shelf supports and full extension drawer slides. European type lower-corner mounted extensions are easiest to install. These are readily available from many building supply stores, or through mail order.

Assemble the base frame first. Figure 14-9 shows how I've used glue blocks to strengthen the assembly. Note the optional miter joint in the outside corners.

Figure 14-7 **Using a jig to make a cope joint**

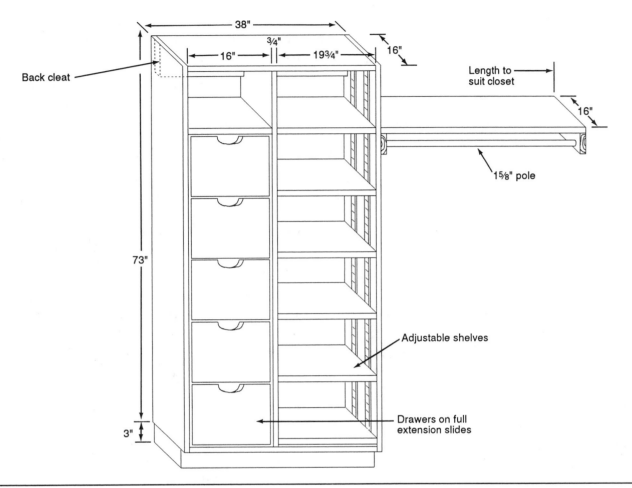

Figure 14-8   Site-built wardrobe unit

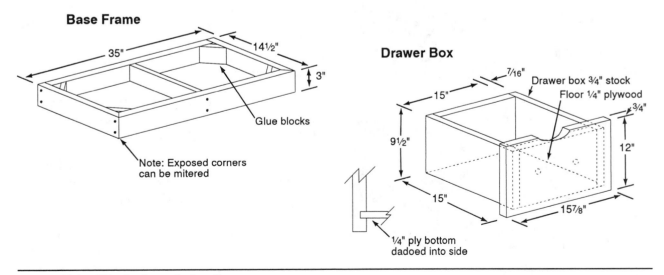

Figure 14-9   Wardrobe unit details

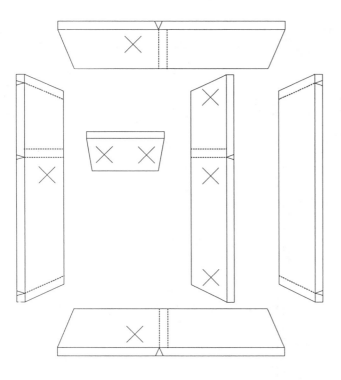

Figure 14-10   **Dry assembly with marks**

Figure 14-11   **Using biscuit joiner with an
                alignment jig**

Set the base into the closet. Shim it level. Then fix it to the floor with 2-inch drywall screws driven at an angle from inside the frame.

Rip the case sides, floor, ceiling, and vertical partition to width. Cut to length. Use the pyramid marking system on the back edge of the stock to identify and match up the components. Check back to Figure 9-6 in Chapter 9 if you need to review this system.

Make marks on the face edge of the floor and on the ceiling to indicate the location of one side of the partition. Mark the face edge of the side components for position of the floor, partition, and ceiling.

Dry assemble the unit using clamps to hold the structure together. Mark an X on the partition, and on the floor and ceiling pieces that face inward. See Figure 14-10. That makes it easier to assemble again and easier to keep from making mistakes when you cut the slots.

Making the slot biscuit joints- Use a simple jig to locate the slots and help hold the tool in position for slotting into the face of the components. Figure 14-11 shows the biscuit joiner layout and position of the jig in use. Note the mark on the cross arm of the jig aligned with the centerline of the joiner base. This indicates the centerline of the slots.

To locate the cuts on the side components for the ceiling and floor, place the arm of the jig at the alignment marks. In this case, they also represent the thickness of the ceiling and floor stock. Then clamp the jig to the stock.

Locate the slots for the partition in both the floor and ceiling components by holding the arm of the jig to the partition alignment mark.

You can use the same jig in a little different way to make the slots in the ends of floor, ceiling, and partition components. Here you hold the component that you're going to end-slot down firmly on a flat surface that's free of any objects or warpage that could make the board rock. The side facing down is the side you marked with an X during the trial assembly.

Bring the cross arm of the jig to the edge of the board. Clamp it securely and run the joiner into the end of the board at the centerline marks. See Figure 14-12. You'll find the side leg of the jig acts as an additional stop for the tool's face locator pins. Keep the tool from sliding sideways as you make the cut.

Figure 14-12   Hold the cross arm of the jig to edge of board for end slotting

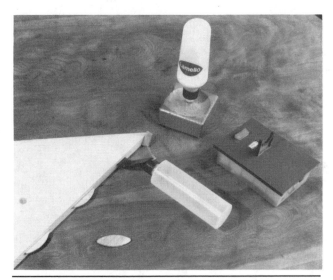

Figure 14-13   Gluing the splines into the slots

Making the drawers- The drawers are simply boxes assembled with spline biscuit joints around a ¼-inch plywood bottom let into a groove.

Glue the splines and the slot for the bottom and tack the box together. Use finishing nails or trim screws to hold the sides in place. Then check for square by measuring diagonally. Set the assembly aside to dry. Install the face to the front of the box with screws from the inside. Make the face 1 inch oversized in length. This lets it cover the ½-inch gap left between the drawer box and case components for the drawer slides.

Assembling the wardrobe unit-
Before assembling the case, mark the position of the drawer slides on one side component and the facing partition wall. If the stock is prefinished, you can install the slides at this time. Mark the position of the shelf standards at the same time. If you choose, you can use a router to make a shallow groove for the standards. This detail lets you cut the shelves within an ⅛ inch or so of the sides.

Now notch for the back cleat in the upper back corner of the partition.

Assemble the unit on a flat surface with the facing edge held up. Insert glue into the slots and press in the splines. See Figure 14-13. A glue bottle made for this handles the job quickly and without a mess.

Hold the components in place with finish nails or drywall screws while the glue sets. Take diagonal measurements to make sure the case is square. When dry, gently flip the unit over on its face and install the cleat to the upper back corner. Install shelf standards and drawer slides, positioning the case so gravity holds the hardware in place while you run in the screws.

Installing the wardrobe- Bring the finished unit to the closet and rest it on the base. Though you've already leveled the base, you might still need to shim between the case and the base frame. When you're satisfied the unit is plumb, screw the case to the base.

Run drywall screws at least 2 inches long through the back cleat and into the wall studs. Figure 14-14 shows how you can cover exposed screw heads with *Titus* caps — available from mail order suppliers.

Figure 14-14   **Cover exposed screw heads with *Titus* caps**

Manhours for Closet Interiors

All figures in the tables are in manhours and are based on the following assumptions:

- Tools and materials needed are available on site.
- The tradesman is a qualified and motivated finish carpenter.
- Work is good quality, stain grade, done no more than 9 feet above floor level.
- All defects are remedied before the carpenter leaves the site.

Add extra time for setup, cleanup, painting or staining, protecting adjacent surfaces, complicated layout or inadequate plans, repair and replacement jobs where fitting and matching is required, working around other trades, setting up scaffolding and ladders for work above 9 feet, and supervision, if necessary. Paint-grade work will usually reduce the time needed by from 20% to 33%.

Closet Poles and Shelving

Including layout and installation.

1¼" closet pole supported with a metal bracket every 32", manhours per linear foot	.04
Add for mitered corner joint, per inside or outside corner	.30
¾" x 12" hat shelf supported with a metal shelf bracket every 32", per linear foot	.10
Add for mitered corner joint, per corner	.30
Site-built closet storage unit with drawers, using plate biscuit joiners and ¾" board or plywood stock, including fabrication, layout, assembly and installation, per square foot of storage unit face	.68
Add for each drawer	.85

Chapter 15

Installing Cabinet Work

Today most cabinets are installed by cabinetmakers or cabinet suppliers. Still, there are times when cabinet installation, or at least part of that job, falls to the finish carpenter on site.

By itself, that's a good reason to know how to install standard cabinetry. Perhaps a better reason, though, is that your local home center or cabinet shop probably subcontracts installation of the cabinets they stock or manufacture. Cabinet installation can be a good, steady source of income for a finish carpenter who knows how to handle this type of work.

I'll guide you through the process step by step. After reading this chapter and doing a few installations, you should feel confident enough to tackle most any cabinet installation job.

Checking the Site and Preparing to Install

Before installing cabinets anywhere in the house, make sure that all the utilities, including ductwork for ventilation systems that will pass through any cabinetry, have been roughed in. Then check the plans. Are you sure the cabinets can be installed as shown? For instance, will installing the cabinets obstruct access to plumbing or electrical outlets? If necessary, arrange to have utility lines moved before you install the cabinets.

The drywall must be installed, taped, sanded, and painted before you begin installing cabinets. If you can, it's best to wait at least a week after the interior finish painting. That way, moisture from the drying paint has a chance to escape from the house. Otherwise, the cabinetwork is going to absorb a lot of this moisture, possibly producing sticking doors and noticeably uneven joints.

You need a few more supplies and accessories to install cabinets than to install trim. Before you leave home to begin a cabinet installation job, review Figure 15-1. You might not need every item on this list, but you'll definitely need the one you didn't bring!

Establishing Level Reference Lines

Before you set the cabinets in place, draw reference lines on the walls to indicate the top of base cabinets and the bottom of wall cabinets. This tells you where to set the cabinets. Establishing these lines is a quick, easy job with a builder's sight level and leveling rod.

If you don't have a builder's sight level, a reservoir-type water level works almost as well. See Figure 15-2. It's a snap to use, inexpensive, and you can do the job solo. Don't bother with water levels that don't have a reservoir — they take two people and are no more accurate.

You can also lay out level lines with a bubble or electronic level attached to a long straightedge. Be sure to adjust the level for

Checklist for Cabinet Installation

Supplies for Installing Cabinets

☐ Countersink plugs of cabinet wood species

☐ Extra lengths of molding stock

☐ Touch-up material to match cabinet finish

☐ Fine steel wool (0000)

☐ Filler putty and caulking material that matches the cabinet finish

☐ Titus screw caps color-matched to wood

☐ Layout plans

☐ Scale ruler

Fasteners

☐ Screws, including:

    ☐ Coarse-thread drywall screws from 1 to 3½ inches

    ☐ Hardware fastening screws

    ☐ System screws in a variety of lengths for 32 mm system hardware

☐ Lag screws for hanging cabinets from the ceiling

☐ Bolts, cabinet-connector type for frameless cabinets

☐ Snap-on clips for removable kickboards

Fittings

☐ Adjustable shelf clips

☐ Extra door and drawer bumpers to replace ones that fall off in shipping and handling

☐ Extra door hinges and pulls

Figure 15-1   **Cabinet installation checklist**

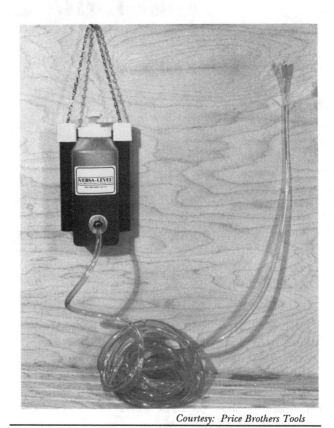

Courtesy: Price Brothers Tools

Figure 15-2   **Reservoir-type water level**

accuracy before using it. This method won't give you quite as accurate a line as the other two methods and it takes longer.

Figure 15-3 shows how to establish reference lines with a water level or a bubble level. Here are the steps you follow to determine these lines:

1) *Using a water level:* Hang the reservoir water level about 40 inches above the floor. It doesn't have to be exactly 40 inches just as long as the distance falls somewhere between the positions of the upper and lower cabinets. Now take the hose around the perimeter of the room and mark the water level on the wall. When you connect the marks, you'll have a perfectly level base reference line for all cabinets.

    Using a bubble level: Establish your base line by attaching a trued level to a long straightedge. Hold the board about

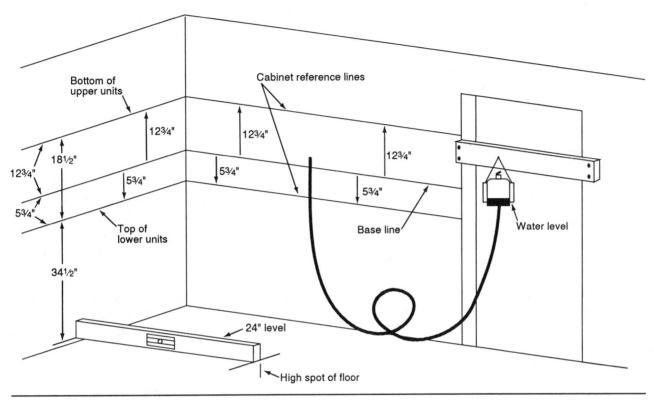

Figure 15-3   **Reference lines for layout of cabinets**

40 inches above the floor, level it, then draw a line along the top edge. Continue the line around the room by extending the line with the straightedge.

Using either a water or bubble level: Mark the location of the wall studs on this base line for future reference when you install the cabinets.

2) Measure down from the base line to the floor at several points. Find the highest point of the floor that you're putting the base cabinets on. Then measure out from the wall 24 inches. Run a level out at this 24 inch point to find where the floor is highest under the cabinets. See Figure 15-3. Nine times out of ten, the highest point on the floor will be well away from the wall.

3) Mark the high point of the floor on the wall. If the high point isn't right next to the wall, transfer the height to the wall with the 24-inch level. See Figure 15-3.

4) Make a mark 34½ inches above this point. Be sure you measure up in a plumb line. This mark represents the top surface of the lower cabinets, assuming a 1½-inch-thick countertop and no additional flooring. This yields a final counter height of 36 inches.

5) Make another mark 18½ inches plumb above this last mark. This is the location of the lower edge of the wall cabinets, assuming a 17-inch clearance between the counter and the uppers. The uppers should fit between this mark and the ceiling or soffit, with room to spare.

6) Measure to the base line from these two marks and note the distance on the wall. Figure 15-3 shows the bottom of the upper units 12¾ inches above the base line. The top of the lower cabinets is 5¾ inches below the line.

7) Establish the two cabinet reference lines. Make marks at 12¾ inches above and 5¾ inches below the base line around the perimeter of the room. Connect the marks with a straightedge. If the base line is level, these reference lines will be level as well.

Once you've got the leveled reference lines in place, refer to the cabinet layout plan and mark the position of the cabinets along the reference lines.

Mark for the base units first. Then plumb up to the upper reference line and mark where you'll align the upper units over the base units. Mark the location and width of any filler strips or spacer moldings. Mark the spaces you'll leave between cabinets for appliances.

Installing the Base Units

Why do I install base units before upper units? Probably out of habit, mostly. Other finish carpenters install the uppers first so the lowers are out of harm's way. If you're using a commercial jack to lift and hold uppers in position, having the lowers out of the way is a definite advantage. However, I often use the lowers as a support when I set the uppers. I tack a temporary counter surface to the lower cabinets and jack the uppers into place using this temporary surface as a brace.

Sometimes cabinet uppers are designed to sit on top of a backsplash that's part of the counter surface. In that case, you have no choice but to put the uppers in after you've installed both the lower units and the counter and backsplash.

Base units can be a breeze to install — if you luck out and get cabinets with European-style plastic leg levelers. See Figure 15-4. These levelers make major height adjustments easy. They also completely eliminate the need to build and level a separate base frame. You don't have to hack away or shim out an integral frame. Instead, a removable toekick board just clips onto the front row of legs that can adjust within a height range of 3 inches to 4½ inches or more.

This removable toekick board also makes it much easier to hunt down and clean up plumbing leaks or to install new flooring.

But let's assume your luck is about the same as mine. We'll level the base frame the hard way.

Setting the Base Frame

Some custom cabinet manufacturers supply a separate base frame for their cabinets. Some even make the frames full length for a run of cabinets. That's a big advantage, nearly as good as having plastic leg levelers

Begin installation of the separate base frame by laying it in place on the floor. The back of the frame should just touch the wall. Be sure it's

Figure 15-4 **European-style plastic leg levelers**

positioned correctly from side to side. Refer to the layout marks you made earlier on the base unit reference line.

Level the base frame this way:

1) Secure the frame to the floor at the high point.

2) Shim wherever necessary to bring the rest of the frame to level.

3) Screw through the frame where shims raise the base above the subfloor.

Setting in the Base Units

On larger units, remove the doors and drawers to reduce the weight. If the cabinets you're working with have clip-on European cup hinges (as shown in Figure 15-5), remove all the

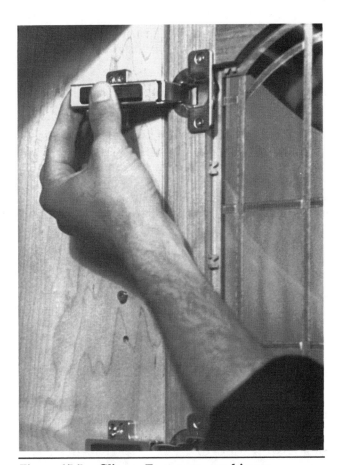

Figure 15-5 **Clip-on European cup hinges**

doors. With these hinges it's easy to reinstall and adjust the doors, so you might as well get the doors out of harm's way and reduce the weight of the cabinet.

Begin by setting each unit in place on its base frame. If the frames are part of the case unit, insert shims under the frames until the top of the unit is even with the reference line. I like to begin with a corner unit and work my way toward open ends or where walls meet.

Keep a careful eye on the reference line layout marks. Be sure you're installing at the correct height and location. Also note where you'll need filler strips between cabinets, and how wide you have to make them. Leaving one of these guys out will really throw you off. Custom-made units require fewer filler strips as a rule, though sometimes decorative moldings will be required between each cabinet.

After you place and align the base cabinets, secure the cabinets together with clamps. Make sure the face frames or the front edges of frameless cabinets are perfectly flush.

I prefer to screw the cabinets to one another before I drive any screws into the wall and the base frame. This gives me a chance to align the front faces of the cabinets perfectly.

Depending on the cabinets, use either screws or special post and screw sets to fix the units together. Attach the units to a separate base frame with 1⅝ inch drywall screws. Cap these screws with Titus fittings. Then attach the cabinets to the wall with 2½-inch drywall screws at each stud. Be sure to insert shims between the back of the cabinets and the wall surface if there's any gap. See Figure 15-6.

Sometimes an appliance which goes under the countertop, such as a dishwasher, will break a run of cabinets, even though the counter may continue across the opening. It's important that you keep the faces of units on each side of the appliance aligned across the break. Before attaching these units to the wall, use a long straightedge to check for alignment of the cabinet base and face.

Figure 15-6   **Shims behind a base cabinet**

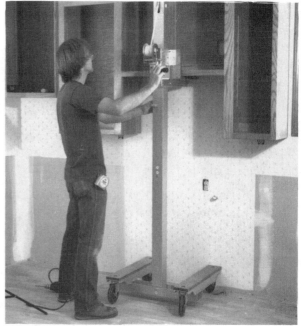

Courtesy: GIL-LIFT Co.

Figure 15-7   **Using a wall cabinet jack**

Installing the Upper Units

Start setting wall cabinets at the corners. Set the bottom edge of the upper units at the reference line you established earlier. Before screwing the unit to the wall, be sure the face and side of the unit are plumb. You may have to insert shims between the back of the cabinet and the wall to get the face plumb.

European-style cabinets usually have a rail and adjustable hanger system for the wall cabinets. In this case, attach the rail level to the wall at the height determined by the matching hardware in the cases. Turn the screws on the hangers to make fine leveling adjustments.

I used to use a couple of automotive-type scissors jacks to hold the cabinets in place while I fussed around with positioning. Now I use a jack made for holding wall cabinets in place. The lift in Figure 15-7 is made by *GIL-LIFT*. This jack can hold several units in position at the same time. Because you join the units before hanging, the chance of getting the faces perfectly flush is excellent. Figure 15-7 shows the lift holding wall cabinets before base cabinets are in. By attaching a special accessory, you can use the lift to hang wall cabinets even after the base cabinets are installed.

As with lower units, check and make sure your fillers show up in the correct locations. Be sure you securely attach the units to one another before you fasten them to the wall. You'll probably have to cut the units over the stove to clear ductwork before you can install them.

Installing Cabinet Components

Once all the cabinets are secured level and plumb, cut the moldings as specified on the plans. Figure 15-8 shows some typical moldings and applied panels used around cabinets.

You may need to cut filler strips to fill a gap between the cabinets and meeting walls. If so, measure the width of the gap every 6 inches along the height of the opening. Transfer these measurements to the filler strip stock at the same intervals. Now connect the marks with a straightedge and cut to the line with a jig saw.

Make the cut with a slight underbevel so the filler strip fits in more easily. (See Figure 10-6 in Chapter 10.) Plane the edge to get a perfect fit.

If your cabinets have adjustable leg levelers, cut removable toekicks for the length required. Use hardwood plywood for these strips. Cut a 5/16 inch saw kerf the length of the strip and about 2 inches down from the top edge of the toekick. Press the clips that attach the kick to the legs into this saw kerf. Then install the strip on the leg levelers.

I like to butt the pieces running out from the wall into the back of the toekicks running across the front of the cabinets. To hide the exposed

end of the toekick, make a return miter (Figure 15-9). Use the same method as shown for window aprons in Chapter 9. Look back to Figure 9-15.

In installations with a rough wood base frame, cover the exposed faces with a run of baseboard stained or painted to match the cabinetry.

With the woodworking phase completed, reinstall any doors or drawers removed from the units. Then install (or reinstall) any fixtures such as lazy Susans and roll-out racks. Make sure these are operating freely and smoothly. Then install any loose adjustable shelving.

If the doors operate with European cup hinges, adjust the hinges so there's an even gap along the door perimeter. Most European-style cabinetry also has hardware that lets you adjust the drawer faces.

Traditional overlay doors and drawer faces on face frame-style cabinets usually don't require fine adjustments for appearance. However, you may have to make adjustments to keep an edge from rubbing on the frame. With many types of hinges, you can make minor adjustments to the hang of a door by opening the door and pushing up or pulling down on it. This bends the hinges slightly and moves the opposite edge of the door up or down. To make larger adjustments, plane the door or drawer rabbets a little deeper. You'll need a special rabbet plane to do this. See Figure 15-10.

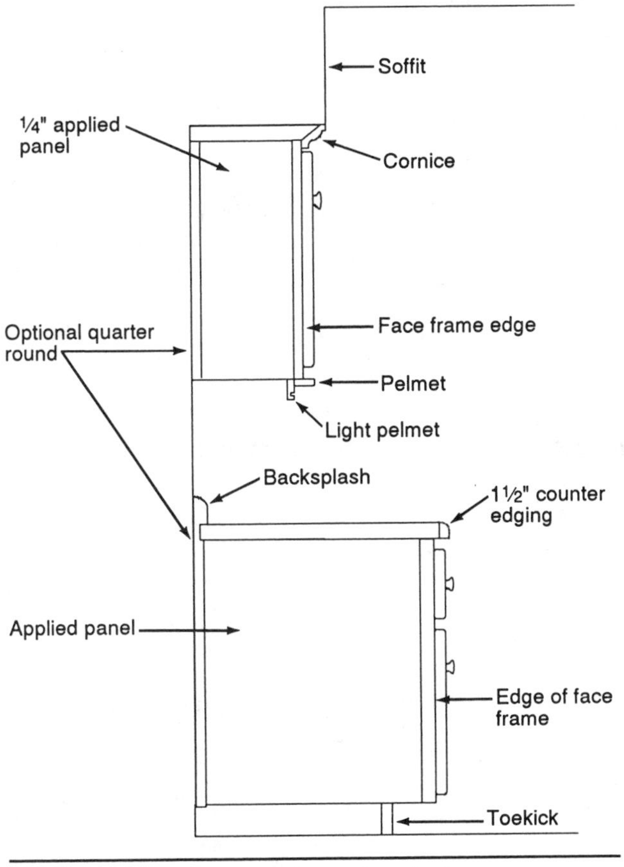

Figure 15-8   **Side elevation of typical kitchen cabinets**

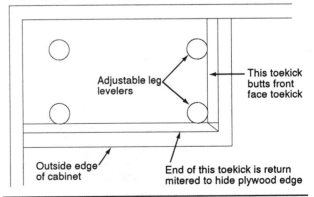

Plan view

Figure 15-9   **Toekick cutting layout**

Your Final Inspection

To leave behind a first class installation, take the time to make a thorough post-installation inspection. Check to be sure you've filled all nail holes. Cap (or in the best work, countersink and plug) all exposed screws. Inspect the finish. Sand out, refinish, and buff out any dings or scratches. Make sure everything that goes with the cabinets is present: fixtures, breadboards, door bumpers, etc.

Finally, vacuum out the interior of the cabinets. Don't leave this job for your customer to do if you want to leave them with a good impression.

Figure 15-10   **A rabbet plane in use trimming back a door rabbet**

Manhours for Installing Cabinet Work

All figures in the tables are in manhours and are based on the following assumptions:

- Tools and materials needed are available on site.
- The tradesman is a qualified and motivated finish carpenter.
- Work is good quality, stain grade, done no more than 9 feet above floor level.
- All defects are remedied before the carpenter leaves the site.

Add extra time for setup, cleanup, painting or staining, protecting adjacent surfaces, complicated layout or inadequate plans, repair and replacement jobs where fitting and matching is required, working around other trades, setting up scaffolding and ladders for work above 9 feet, and supervision, if necessary. Paint-grade work will usually reduce the time needed by from 20% to 33%.

Cabinet Installation

Including layout, leveling and installation. No fabrication, assembly or countertops included. Manhours per cabinet

Base cabinets

Cabinets to 36" wide	.30
Cabinets over 36" wide	.35

Wall cabinets

Cabinets to 18" wide	.25
Cabinets over 18" to 36" wide	.30
Cabinets over 36" wide	.35
Oven or utility cabinets, 70" high, to 30" wide	.45
Filler strip between cabinets, measured, cut, fit and installed, per each	.20

Chapter 16

Fireplace Surrounds

When a fireplace is built into a wall, it leaves a gap between the masonry and the wall framing. Covering this gap is about as much fun as a finish carpenter gets these days. It's a job that includes creative design work and cutting and fitting a wide variety of moldings. Figure 16-1 shows the traditional way to cover this joint.

Of course, you can buy this collection of molding (called in the trade the *mantel and surround*) prefabricated at the mill. But it's still a challenge, and a bit of fun, to install it properly. This chapter will deal first with the site-fabrication of a surround and mantelpiece, and then with the installation of a prefabricated unit.

Surround Moldings

The surround molding has two basic elements:

1) Vertical pilasters that form the two sides

2) A horizontal frieze board that serves as a cap for the pilasters

These boards are usually "made fancy" with the addition of moldings and routed surfaces. There may be a plinth at the base of the pilasters, and back moldings between the pilasters and the wall and the masonry. You can work coves or straight grooves into the pilasters to give the

impression of fluted columns. Finally, there's a variety of band and bed moldings you can use to trim out the frieze board.

The mantel sits on the top edge of the frieze. Its overhang is supported either with corbels or with a crown or dentil molding. The mantel's edge is often decorated with band moldings or a routed shape. The drawing in Figure 16-2 illustrates the parts of a typical installation.

Design Factors

When designing the unit, check what your local code requires for distance between the fireplace opening and the molding that surrounds the opening. Generally, you can't install any combustible material within 12 inches of the fireplace opening. Also note how far away the mantel has to be from the opening. With most codes this depends on the distance the mantel projects out from the wall. The clearance requirements for masonry fireplaces are listed in Section 3707 of the *Uniform Building Code.*

Attaching the mantel to masonry can be a real hassle. That's why I try to design the surround so the mantel starts where the wall framing begins above the fireplace. There'll be times, however, when the design of the masonry won't give you that choice. In that case, you'll have to attach the mantel to wooden plugs or lead anchors you set into the masonry.

If the masonry projects out from the surface of the wall, you'll have to add trim to the pilasters and frieze board to join them back to the wall.

Courtesy: Maizefield Mantels

Figure 16-1   Completed mantel and surround

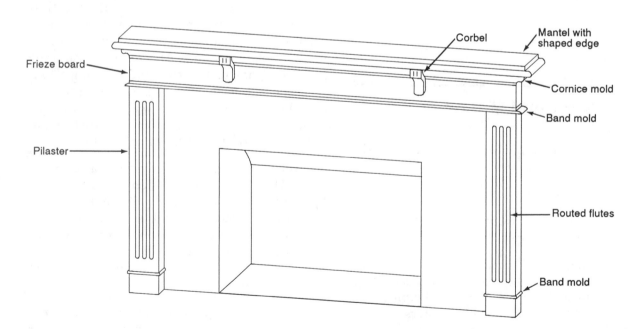

Figure 16-2   Traditional surround and mantel components

Flush opening

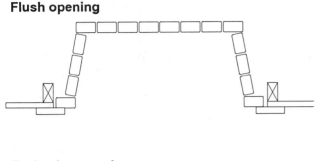

Projecting opening

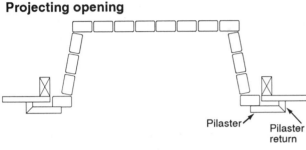

Pilaster

Pilaster
return

Figure 16-3   **Plan view of fireplace opening**

Figure 16-3 shows how to handle a projecting fireplace. For best appearances, join the edges of the return boards to the pilasters and to the ends of the frieze board with a miter joint.

Make the pilasters and frieze board from solid wood for stain grade work. For paint grade you can make the pilasters and frieze board from sheet stock such as plywood or MDF board. Choose MDF over plywood if you're going to shape the surfaces or edges with a router. MDF also provides a much smoother surface that's easier to paint.

Assembling the Surround

You need a large flat surface to lay out and join parts of the surround. I recommend tacking a clean sheet of ¾-inch plywood to the subfloor in an area close to the fireplace. Sweep up the subfloor first so the plywood lies perfectly flat.

Here are the steps to follow to assemble the surround:

1) Cut the frieze and pilaster boards to size. Rout in any decorative shapes you've decided on. If the ends of the frieze

board and the outside edge of the pilasters must be mitered, cut them now. Be sure to leave the pilasters at least an inch long to allow for final fitting.

2) Now lay the parts out on the plywood. Some carpenters make the frieze and pilasters out of one piece of ¾-inch hardwood plywood. I've always looked on this as a waste of materials — what do you do with the piece you cut out? And it doesn't really save you any time.

   There's one advantage to plywood, though. It won't shrink when exposed to heat from the fireplace. On the other hand, shrinkage isn't really a problem if you design the moldings to hide the effects of movement.

   If you choose to use plywood, skip Step 3.

3) Join the pilasters to the frieze board using dowels or spline biscuits. Make sure the parts are square to one another. Stop assembly here until the glue is thoroughly dry.

4) If return boards are needed, cut and fit them to the assembled surround. Cut them wider than necessary so you've got some room to scribe them to the wall.

5) Attach the band and bed moldings to the surface of the surround. Don't attach returns that join the moldings back to the wall just yet. Wait until after you install the surround.

6) If corbels are used to support the mantel, attach them to the frieze board. Be sure to locate them square and flush to the top edge of the frieze. Attach them with glue and screws from the back.

Installing the Surround to the Fireplace Opening

When the glue has set, stand the assembly up and hold it in place in the opening. If you cut the pilasters longer than needed, this is the time you trim them. It's also your chance to get the top of the frieze board perfectly level. Here's how:

1) Place a level on the top edge of the frieze board. If you're worried about the level falling off — and it probably will — then tape it to the frieze or use double-stick tape between the level and the board.

2) If the frieze isn't perfectly level, insert shims under one pilaster until it is.

3) Use a pencil compass to scribe to the distance you want to move the frieze board down. Scribe this distance from the floor onto the bottom of the two pilasters. See Figure 16-4.

Remove the assembly from the opening and cut the pilasters to the scribed length. Now set the surround back in the opening.

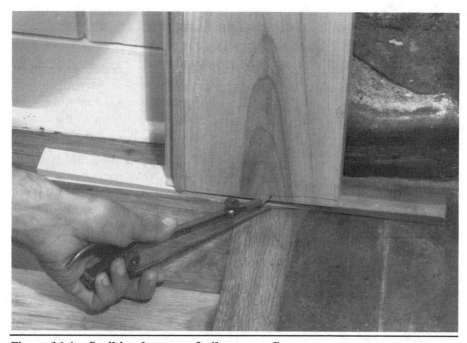

Figure 16-4   **Scribing bottom of pilasters to floor**

If the masonry projects out from the wall, you'll need to cut the surround returns to fit. Holding the face of the surround plumb, shim between the wall and the surround if necessary. Then scribe to the wall and cut to your scribe line. A slight underbevel of 2 to 3 degrees literally lets the edge cut into the wall surface, making a perfect fit.

To install the surround to the opening, set finish nails or screws through the boards and into the wall studs. When possible, place the fasteners where they'll be hidden or covered by the moldings. Finally, cut and install the returns for the band moldings.

Installing the Mantel

Cut the mantel to length. Then cut the width slightly wider than you need so there's room to scribe it to the wall. The next step is to rout or shape the edge molding. Do whatever detailing is required.

When you've finished detailing, find the center point of the mantel and make a mark on the bottom surface. Mark the centerline of the surround on the top edge of the frieze board.

If the surround doesn't have corbels to support the mantel, add small metal L brackets. They'll be hidden behind the crown molding. Or you can set the mantel on a 2 x 4 installed to the wall just over the frieze board. Figure 16-5 shows all three options.

Place the mantel on the surround, making sure the centerlines are even with one another.

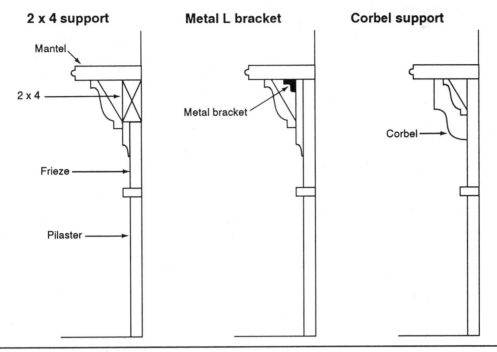

Figure 16-5   **Cross section of frieze and mantel**

Here's how to scribe the mantel to the wall:

1) First make sure the overhang of the mantel past the face of the frieze board is the same at each end.

2) Set your pencil compass scribe for the maximum span the mantel must move to close any gap between it and the wall. Add an eighth of an inch to prevent the pencil from catching on the edge. This is similar to fitting wall paneling. Review Chapter 12 if you need to refresh your memory.

3) Draw the scribe line, making sure you hold the compass at a right angle to the front edge of the mantel. As I warned in Chapter 12 (Figure 12-12), curb that tendency to hold the scribe perpendicular to the surface of the wall!

Remove the mantel from the surround frame. Cut on the scribe line with a jigsaw or circular saw. Now set the mantel back on the frieze and carefully fit the cut with a block plane. It should fit tight to the wall surface and evenly overhang the frieze board along its entire length.

When you're satisfied with the fit, secure the mantel to the surround frame and the wall studs. If you have to attach it to masonry, drill through the mantel into the masonry with a ⅛-inch drill bit at the location of the fasteners. You won't have much of a drill bit left, but you'll have clear marks where to place the wood or lead anchors. If possible, drive anchors into the mortar, not the bricks or stones.

If you can't hit the mortar, locate the anchor as close to the middle of the brick or stone as possible. Otherwise inserting the fastener and expanding the anchor may split the masonry.

Trim out the junction of the mantel and the frieze board with cornice, bed, or dentil moldings.

Finish up by plugging exposed countersunk screw holes with wood bungs. Fill any nail holes with putty. Shape and sand the joints of the moldings.

Installing a Prefabricated Unit

You may not fabricate every mantel and surround, but you're almost certainly going to do the installation. If so, I recommend that you offer some suggestions on design of the unit. At the very least, be sure to review overall measurements before ordering the unit. Figure 16-6 shows the dimensions you'll have to know before ordering:

■ The length (A) and width (B) of the mantel shelf.

■ Overall width of the body (C) and height of the unit (D). The body measurement is taken to the outside edge of the pilasters. The height is measured to the top of the mantel.

■ Width (E) and height (F) of the opening. Most manufacturers suggest that the distance between the open firebox and the inside edge of the pilasters and frieze board be at least 6 inches. This meets the code requirements.

■ The projection for the facing (G). Some manufacturers may call this the depth of the cavity. This measurement depends on how the masonry sits relative to the wall surface. Unless they're perfectly flush — or close enough so a piece of back molding or quarter round will cover any gap — the unit needs returns that can be scribed to the wall and the fireplace material.

When measuring for the projection, be sure to locate the point where the masonry projects the farthest into the wall. Also measure and account for any out-of-plumb in the wall surface.

Installing a prefabricated unit is a lot like installing a unit you made. The only difference is that the units usually have the mantel preattached to the surround. In this case, unless the mantel has been glued to the frieze, remove it and install it separately

If it has been glued, you'll have to scribe the mantel and surround to the wall and masonry in one step. Give yourself some leeway when cutting to the scribe line. Notice that it's impossible to make the cut entirely with the jigsaw or circular saw. The overhang of the sides of the shelf prevents this. So finish it up with a handsaw.

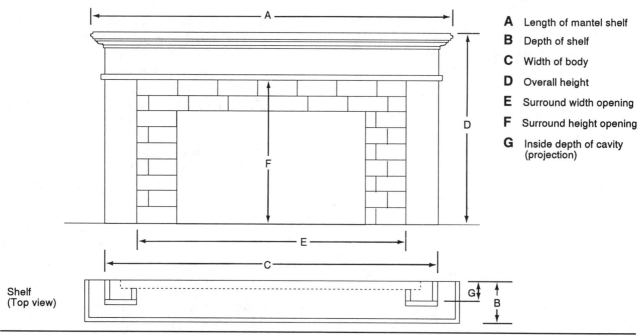

A   Length of mantel shelf

B   Depth of shelf

C   Width of body

D   Overall height

E   Surround width opening

F   Surround height opening

G   Inside depth of cavity (projection)

Shelf (Top view)

Figure 16-6   **Size specifications for a prefab mantel/surround unit**

Depending on the situation and your personal tool preference, it might be easier to trim back to the cut line with a belt sander or grinder fitted with a sanding disc. Otherwise, use a block or jack plane.

Constantly check the fit of the unit to the wall as you're trimming it to the scribe line. When you're satisfied with the fit, install and finish out the unit as described earlier for site-built models.

Manhours for Installing Fireplace Surrounds

All figures in the tables are in manhours and are based on the following assumptions:

- Tools and materials needed are available on site.
- The tradesman is a qualified and motivated finish carpenter.
- Work is good quality, stain grade, done no more than 9 feet above floor level.
- All defects are remedied before the carpenter leaves the site.

Add extra time for setup, cleanup, painting or staining, protecting adjacent surfaces, complicated layout or inadequate plans, repair and replacement jobs where fitting and matching is required, working around other trades, setting up scaffolding and ladders for work above 9 feet, and supervision, if necessary. Paint-grade work will usually reduce the time needed by from 20% to 33%.

Fireplace Surrounds and Mantels

More time will be required when a mantel or molding must be scribed around irregular masonry surfaces.

Layout, cutting, fitting, assembling and installing a vertical pilaster on two sides and a horizontal frieze board above the pilasters. Any additional moldings will take more time. Manhours per linear foot of pilaster and frieze 35

Mantel shelf attached to masonry, including cutting,
fitting & installation, per mantel 75

Installing a prefabricated mantel and fireplace surround, per each 1.25

Chapter 17

Flooring

Laying wood flooring is no piece of cake. If you're not feeling the pain in your knees, it's because your back is hurting too much. But it's rewarding work for a good finish carpenter. A job well done makes a long-lasting impression both on a home and on its owner.

Applying the finish floor is one of the last tasks on a job site. The only finish carpentry left after that is installing baseboard and base cabinets in rooms where flooring is being laid.

I recommend that you install finish floor wall-to-wall, under all cabinets. It adds a little to the flooring material cost. But there's a lot of labor saved in not having to cut and fit around casework. Another advantage to laying flooring under the cabinets comes to light if the home is remodeled. When the cabinet placement changes, flooring is already in place everywhere it's needed.

And finally, when you install cabinets last, you improve significantly the chance that those delicate door and drawer faces will survive the flooring process undamaged.

Flooring Types

In this chapter I'll cover the installation of two common types of finish floors that carpenters are asked to install:

1)   Tongue and groove strip flooring

2)   The wider straight-edged plank flooring

Installation technique for other types of wood floors, such as blocks or individual parquet, depends on the brand of flooring. Your best bet is to follow the manufacturer's instructions.

The most common tongue and groove strips are ¾ inch thick by 2 inches wide. But you can get thinner and wider stocks. Most hardwood strip flooring is also end-matched, meaning that one end has a tongue and the other has a groove. You'll seldom find fir or other softwood flooring end matched. Side edges are commonly underbeveled to assure a tight edge-to-edge fit. Undersides are ploughed out with shallow grooves to cut down on warpage and to ensure that the strips lay flat.

Plank flooring usually isn't end matched, side beveled or back ploughed and may even lack tongues and grooves. The installer may have to cut the grooves and install splines. Straight-edged boards are usually ¾ inch thick and may be up to 11 inches wide. Use these wider planks to create a traditional colonial-era floor.

Woods for Flooring

The most common wood used for flooring is red oak. It's almost always sold as tongue and groove (T&G) stock. Other species milled for strip flooring include hard maple, white oak, and tan oak. Cherry and hickory are also available.

Spruce and pine are sometimes sold as tongue and groove *decking*. This material has disadvantages when used as finish flooring:

1) It's rarely dried to anywhere near the proper moisture content for interior use. That makes it certain to shrink and warp.

2) The chamfered edges of tongue and groove decking catch a lot of dirt.

Hardwood plank flooring tends to expand and contract with changes in moisture content. Choose the wood species with caution. The more unstable the species, the narrower the planks must be to avoid problems.

Softwood plank flooring doesn't expand and contract as much as hardwood. You can use wider planking without the shrinkage problems common with hardwood. Figure 17-1 lists characteristics of woods used in flooring.

Preparing for the Flooring Job

To avoid callbacks on wood floor jobs, start with good flooring material. Inspect the wood to see that it's relatively straight and flat, free of major defects, and as close to the moisture content level of the house as possible.

Softwoods				
Species	**Resistance to wear and abrasion**	**Freedom from shrinking and swelling**	**Resistance to splitting under fastenings**	**Ability to provide a smooth finished surface**
Southern yellow pine	High	High	Medium	High
Spruce	Low	Medium	Medium	Low
White pine	Low	High	High	Medium
Douglas fir	Medium	Medium	Medium	Medium
Redwood	Medium	High	Medium	Medium
Hardwoods				
Red oak	High	Low	Medium	High
White oak	High	Low	Medium	High
Maple	High	Low	Low	High
Cherry	High	Medium	Low	High
Ash	High	Medium	Medium	High

Figure 17-1   **Characteristics of wood commonly used for finish flooring**

Specialized floor-laying tools
1.   Moisture meter

2.   Power tongue nailer
    a.   Hand operated
    b.   Pneumatic

3.   Power face nailer (optional: a pneumatic finish nailer is an excellent alternative)
4.   Power or hand-operated jamb saw (for cutting back the base of standing moldings so flooring can slip under it)
5.   Rubber hammer (to set off power nailers and to knock flooring strips into place)
6.   Draw chisel or Wonderbar®
7.   Slotter/router table (for end-slotting flooring)
8.   Drum sander
9.   Floor polisher (optional: for surface sanding at finer grits)
10.   Circular edge sander
11.   Random orbit sander (for sanding out swirl marks at junction of drum and circular sanders)
12.   Floor scraper (for surface flooring in corners of room)
13.   Trowel (for application of filler)

Figure 17-2   **Tools for the installation and surfacing of wood floors**

To be sure the material will be in good shape when it's installed:

Have the wood delivered to the closed-in and heated home (at least 65 degrees) no less than one week prior to installation. Three weeks is even better. This gives the wood time to reach the moisture level of its new home.

Don't heat the work space with portable propane heaters. They give off a tremendous amount of moisture. This defeats the purpose of trying to bring the moisture content down to what's normal in a home.

Allow at least three days for interior paint to dry. Latex paint needs four days. Any concrete poured inside the home should also be thoroughly dry.

Stack the flooring away from heating ducts. Unbundle the strips and distribute them at random so air can circulate around each piece. To speed drying, layer the boards with ½ inch-thick strips of wood in between each course.

While unbundling the piles, check out the boards as you go along. Cull out any defective stock. You'll either trash it or use it in less visible areas, such as closets.

Specialized Floor-Laying Tools

You probably have most of the hand tools needed for cutting and fitting flooring. But some more specialized floor installation and finishing tools will help increase productivity.

Figure 17-2 lists some of these more specialized tools and describes how they're used.

Inspecting and Preparing the Subfloor

Check the type and thickness of the subfloor to be sure it's adequate for the kind of flooring you're installing. Figure 17-3 lists the underlayment choices for each type of flooring.

Note that underlayment needs to be heavier if flooring runs in the same direction as the floor joists. Also note that there's no particleboard in this table. That's because particleboard absorbs and retains moisture. I never use it under a wood floor. Also, it doesn't hold fasteners very well.

If the size and type of the underlayment is adequate, continue your inspection by running through this checklist:

Moisture content- Laying wood flooring with a high moisture content is almost always a mistake. The wood will shrink as it dries, opening gaps at the joints and splitting wood where nails are driven. When that happens, your reputation is bound to suffer. Warn the owner or the owner's

Type of flooring	Underlayment (choose one)					
	¾" wood planking laid at diagonal	2 layers of ½" plywood. Staggered seams	¾" T&G plywood	1" T&G plywood	2 layers of ⅝" plywood. Staggered seams	1⅛" T&G plywood
Run perpendicular to joist direction						
Random length T&G strip flooring	✓	✓	✓			
Shorts (under 18") T&G strip flooring		✓		✓		
Plank flooring (without tongues or splines)					✓	✓
Run parallel to joist direction						
Strip or plank flooring					✓	✓

Figure 17-3   **Recommended underlayments for a variety of finish flooring**

representative about the problem. If they still want you to go ahead with the job, have them sign a waiver of responsibility.

Use an electronic moisture meter to take moisture content readings at several places on the underlayment, especially in the vicinity of exterior doors. In most parts of the U.S., the moisture content should be less than 12 percent. If in doubt, get the opinion of your local flooring supplier on the correct moisture level.

Also test the moisture level of the subflooring. A wet subfloor will release moisture into drier flooring laid on it, making the finish floor expand and cup. Don't rely on Kraft paper placed between the subfloor and finish floor to stop the transfer of moisture. It only reduces the rate of exchange.

Also check the moisture in the joists. It should also be less than 12 percent. If not, shrinking joists can eventually cause the floor boards to separate or squeak.

Loose edges- Walk around the perimeter of each underlayment panel. Listen for squeaks and watch for movement. Nail down loose areas with ring shank nails. Gluing underlayment to the joists with construction adhesive creates a much better bond with the joists.

Unevenness- Use a long straightedge or mason's string to check for humps and hollows in the floor. You'll usually find them where the underlayment panels meet. Sand out the high spots with coarse paper on a belt or drum sander. Be sure to set the nails in the area to be sanded so you don't rip the sandpaper. Build up low spots by cutting layers of resin paper to the shape of the hollow. Then tack these layers in place.

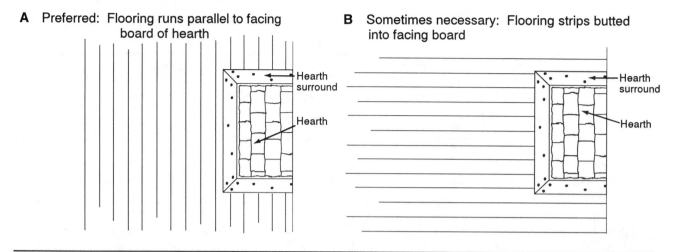

A Preferred: Flooring runs parallel to facing board of hearth

— Hearth surround

— Hearth

B Sometimes necessary: Flooring strips butted into facing board

— Hearth surround

— Hearth

Figure 17-4 **Direction of flooring strips at hearth surround**

Blocking- Inspect the underlayment from below. There should be blocking under all the seams of the panels. If not, cut 2 x 4s to length and install them between the joists. Nail off the panel joints from above.

Insulation- While you're still under the floor, check to see that there's plenty of insulation between the underlayment and the furnace and heat ducts. Hot spots in the floor cause uneven shrinkage. Put your hand on the insulation at random spots and feel if it's dry. If it isn't, have it replaced.

Moisture again- While you're in the basement, check for any standing water. If you see puddles, something isn't draining right. Get someone to solve that problem before you start laying the floor. If it goes unsolved, it's responsibility waiver time again.

Laying Out the Floor

Most finish carpenters prefer to have the length of the strips run parallel with the long dimension of a room. There are some good reasons for this type of layout:

■ It makes the room appear larger.

■ An occasional short strip isn't as noticeable in a long run.

■ Shrinkage and expansion occur across the width of the strips, not the length. The shorter the width, the less expansion and contraction with changes in humidity.

■ You'll be laying strips across the joists and can nail each strip through the underlayment to an underlying joist.

Figure 17-4A shows a plan view with a fireplace at the end of a room. Six-inch-wide boards have been mitered to form a plank surround for the masonry hearth. The flooring is laid parallel with the front of the hearth. Note the notched starting board. I think this looks better than Figure 17-4B, where the flooring is perpendicular to the hearth, butting into the plank surround.

But sometimes structural considerations are more important than appearance. If the underlayment isn't heavy enough, the flooring has to run *across* the floor joists, which may result in the view shown in Figure 17-4B. Otherwise you'll risk getting a floor that squeaks. Without adequate support, the floor constantly flexes under use, loosening the fasteners. Squeaks happen when grooves rub against tongues that no longer fit snugly.

When flooring layout changes from room to room, plan to locate the joint under a door. Another option is to install a plank across the door opening and butt the flooring to it from both rooms.

At stair landings, run the flooring up against a nosing strip. Make sure you know ahead of time what the width of this strip will be.

Figure 17-5 shows a typical flooring layout. Note that the strips in the largest room of the house parallel the face of the hearth surround. The direction of the flooring in this room influences the direction the flooring runs throughout the rest of the house. The exception is the room in the southeast corner where, for the sake of appearance, the strips are laid the long dimension of the room. Because the plank across the door opening separates the flooring, the change in direction isn't obvious.

Setting Starting Course Baselines

Once you've decided on the layout, the next step is to decide where to lay the starting course. I like to start in the largest room in the house and get the bulk of the job done while I'm still fired up for it. It's also easier to use the off-cuts and culls in smaller rooms.

Where you start in that room depends on how accurate the framers were. Measure the distance between the opposite walls at both ends of the room. If the opposing walls are within ½ inch of being parallel, you can start on either side of the room. If they're more than ½ inch off parallel, lay the starting course parallel to the most visible wall. If both walls are highly visible along the length of their base, you'll have to compromise. See number 3 in the steps on the next page for establishing the baseline.

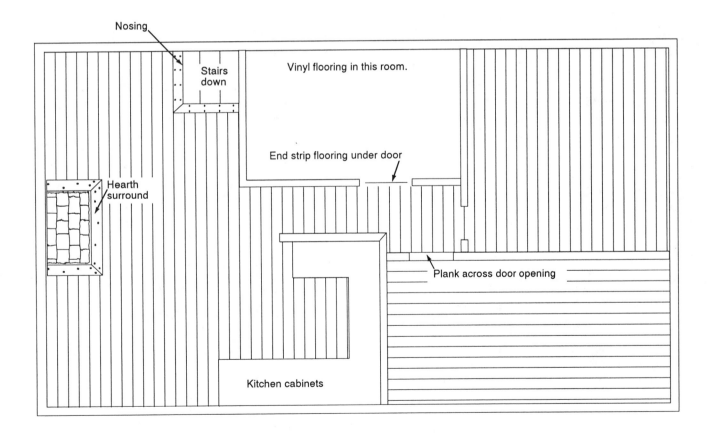

Figure 17-5   Typical flooring layout

If the room has boards surrounding a fireplace hearth, lay the first course against the front plank. I'll talk about how to do this a little later.

Try to install the first course of flooring as straight as possible. Follow the steps below to create a baseline to indicate the position of the front edge of the course:

1) Lay a strip of paper vapor barrier along the wall where your starting course will go. I use three-layer resin-impregnated Kraft paper. Staple it down.

2) At one end of the room, measure out from the wall the width of one strip of flooring plus an expansion gap. This expansion gap should be one-half the distance you expect the flooring to expand in humid weather. For most hardwood flooring, allow an expansion gap of $\frac{1}{16}$ inch per foot of flooring width. Thus, a 12-foot width of flooring might expand $\frac{12}{16}$ inch ($\frac{3}{4}$ inch). In this case, leave a $\frac{3}{8}$ inch expansion gap along the first wall. Make a mark at this point on the paper. The other half of the expansion gap ($\frac{3}{8}$ inch in this case) goes along the opposing wall.

3) If the walls aren't parallel and the opposing wall won't be hidden at the baseboard line, you'll need to compromise. In this case, measure how much the walls are out of parallel and add half this amount to one of the baseline marks. See Figure 17-6. In the illustration, the wall at the top of the drawing is 2 inches longer than the one at the bottom. To compromise the line, I added half of the 2 inches (1 inch) to the top measurement away from the starting wall.

4) Snap a chalk line connecting the marks at either end of the room. Lay the front edge of the first course of flooring on this line.

A final note about expansion gaps. There'll be times when the expansion gap needed is too large to be covered by the baseboard. That's

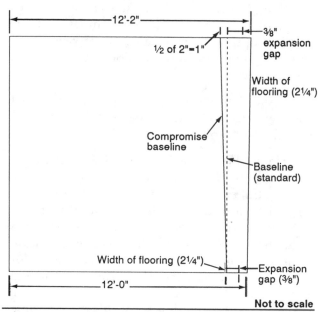

Figure 17-6   **Compromise baseline for out-of-parallel opposing walls**

especially true if no base shoe is used. In that case, borrow a little space from the drywall. Either hold the drywall $\frac{3}{4}$ inch above the floor line or cut out the bottom $\frac{3}{4}$ inch of drywall at the base of the wall. This provides extra room for the expansion gap on both sides of the room.

If you foresee a problem with the gap, give the drywall crew some scraps of flooring. Ask them to hold their drywall up above the floor on the scraps you gave them.

Installing Tongue and Groove Strip Flooring

Start like this:

1) Mark the location of the joist centerlines on the bottom edges of the walls.

2) Lay a vapor barrier of 3-ply resin over the underlayment. Overlap the edges by at least 4 inches. Be sure to overlap

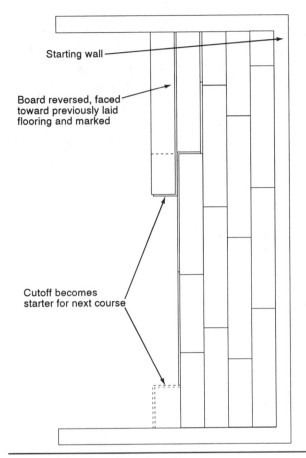

Starting wall

Board reversed, faced toward previously laid flooring and marked

Cutoff becomes starter for next course

Figure 17-7   **Starting piece for new row**

toward the wall where the first course is laid. This keeps the strips from catching on the overlapping edges of the paper as you slide the boards into place.

3) Mark joist centerlines on the paper to make nailing easier. Hold the chalk line at the joist centerline marks you made on walls. Then snap a centerline across the room on the building paper. Continue to mark the centerline of each floor joist.

4) Install any wood strips that surround hearths, heat ducts, and other protrusions.

5) Install nosings at stairway landings and boards across door openings. Don't use temporary sticks for the nosings or door

boards. There's a good chance these will distort under pressure from the flooring, creating gaps when you place the permanent fittings.

Install the first course of flooring to the baseline like this:

1) Select the straightest stock you can find for the first two or three courses. You need a good, straight start for the rest of the runs.

2) Lay the first board to the right as you face the starting wall. The edge and end tongues will be facing out. Hold the board slightly away from the end wall; up to ½ inch is fine if baseboard will cover the gap. Align the front edge to the baseline.

3) You'll be working close to the wall for the first three or four rows. Prevent possible damage to the walls by using an air-driven finish nailer rather than a power nailer to start. Face nail the piece to the floor along the back edge where the nail will be covered by base shoe. If there won't be a base shoe, plan on filling the holes later. Nail through the tongue at a 45-degree angle. If you're laying the flooring across the joists, drive one pair of nails into the joists and another pair midway between joists.

4) Facing the starting wall, work from right to left. Tap the strips into place with a rubber mallet by hitting the end and edge tongues. When it's time to fit the last piece to the far end wall, select a piece about 12 inches longer than the length to be filled. Mark the length by reversing it so the tongue faces the starting wall. Hold it back from the end wall (up to ½ inch) and make a mark where it crosses the end of the last installed piece. Cross cut the piece at the mark and install it. Use the cutoff to begin the next row. See Figure 17-7.

Figure 17-8   Using a Japanese saw to cut back
jamb molding to height of flooring

Figure 17-9   Air-driven power nailer

Installing a First Course to a Hearth Surround

If the floor has planks surrounding a hearth or other protrusion, lay the first course parallel to the plank along the front edge. For the sake of appearance, notch the floor board around the mitered corners of the hearth surround, as shown back in Figure 17-4A.

Establish the baseline in this way:

1) Measure out at each end of the front plank the width of the flooring. Subtract for the depth of the notch.

2) Connect the marks with a straightedge. Then extend the line with a string to either end of the room.

3) Lay the front edge of the starting course to this line, cutting the notch to fit.

Continue to lay flooring back toward the wall on either side of the hearth after inserting a spline or *slip tongue* in the groove of the starting course. This turns the course into a double starter, allowing you to run the flooring strips tongue-side-out in either direction from this starter course. If your flooring supplier can't provide you with a length of slip tongue, rip one

out of a length of oak. Be sure it fits snugly in the groove, and protrudes out no more than the flooring's standard tongue.

Installing the Middle Rows

Continue installing rows across the width of the room, tapping the pieces into place with a rubber mallet. If a strip won't fit snugly against the adjacent strip, check for obstructions hiding under the tongue or in the groove. To get a warped strip into place, use a pry bar or a draw chisel as a lever. Place a scrap of flooring between the lever and the strip to protect the edge.

Run the flooring under, not butted to, standing moldings. To cut standing molding to the correct height, use a scrap of flooring set under a Japanese saw or a special flooring saw as a guide. Then saw through the molding, as shown in Figure 17-8.

Use a power nailer to fasten strips to the floor. If applying strips across the joists, nail through the tongue at each joist and in the center of the span between joists. Figure 17-9 shows an air-driven power nailer that holds a nail at the tongue and drives it at the correct angle

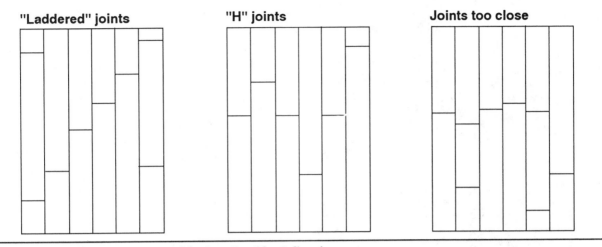

"Laddered" joints   **"H" joints**   **Joints too close**

Figure 17-10   **Three layouts to avoid when "racking" flooring**

automatically. Note also the rubber hammer in Figure 17-9. It's used to both tap flooring pieces into position and to set off the air-driven power nailer.

Keep four or five courses laid out ahead of the work. This helps you see in advance where butt joints of adjacent courses will fall. For the sake of appearance and to minimize squeaks,

stagger joints in adjacent courses by at least 3 or 4 inches. I also avoid staggering joints in a sequence or having butt joints directly opposite one another separated by one course. Figure 17-10 shows examples of these problems. Also, avoid grouping too many short boards in one area of the floor.

When you've got several rows of flooring laid out ahead of the work, nailing strips is quick and easy with a power nailer. See Figure 17-11.

Installing the End Rows

You can't nail the tongue of the last three or four rows approaching the opposing wall with a power nailer. So switch back to the air-driven finish nail gun once again. Nail through the tongues for as long as possible. Face nail only the last two rows.

It's Murphy's Law that the last row will never be exactly the width of a board. The last strips will have to be ripped to fit. If the last row would be less than about 1 inch wide, rip both of the last two rows so

Figure 17-11   **A power nailer makes the job quick and easy**

Figure 17-12 **Using a draw chisel to snug up the flooring**

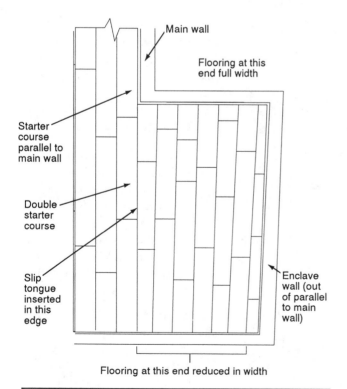

Figure 17-13 **Tapering courses to meet out-of-parallel wall**

they're roughly the same width. This won't be necessary if you've got a selection of wider flooring strips on hand. In general, a wider last strip looks better than several narrow strips.

In any case, measure the gap at every other joist before installing the last two rows. Rip lengths of flooring to fit. Use a draw chisel to pull the strips tight to the last course installed. Figure 17-12 shows the procedure. I use the finish nailer to face nail the last strip while the chisel holds the strip tightly in place. Notice that the chisel is levering off the *floor* under the drywall, not the drywall itself.

Tapering to fit- Many homes have an enclave — a portion of the room that extends beyond the rectangular perimeter. Laying strips in the enclave doesn't present any special problems unless the enclave wall isn't parallel to the opposite wall. Unfortunately, that's often the case. See Figure 17-13.

The final row of strips along the enclave wall has to run parallel with that wall. You do this by planing a taper into the groove side of each row of flooring pieces, beginning at least four to five courses away from the wall. Figure 17-13 shows the section of flooring where strip widths have been reduced. If you remove more than 1/8 inch in any strip, plane down the tongue of the joining piece. Otherwise the tongue might hold the adjoining strip away, producing a gap.

Installing a Plank Floor

In most ways, laying plank floor is the same as laying tongue and groove strip floor. But there are a few differences worth noting.

Start by inspecting the underlayment. Note that plank floors require a thicker underlayment if they aren't going to be splined. Follow the same layout and baseline procedures.

Plank flooring should be either tongue and groove or slotted and splined along both the plank's edges and ends. This keeps the floor level and keeps the planks from shifting.

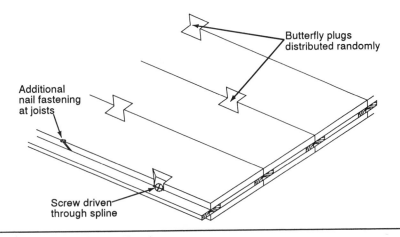

Figure 17-14 **Butterfly plugs**

Attach planks to the floor joists the same way you do strip flooring. If the planks are wider than 4 inches, it's generally best to face screw the boards through to the joists. Screws should be countersunk and plugged. I often use a contrasting wood for the plugs for a decorative effect.

Figure 17-14 shows a second method of attaching planks — butterfly plugs installed over screws run through the splines. The advantage to this method is that the planks can shrink without splitting around the screws. First, cut a recess for the butterfly plug to the depth of the spline. Run a screw through the spline. Then inlay a butterfly plug to hide the screw. Butterfly plugs look best when scattered randomly over the floor surface. Give the flooring some additional staying power by nailing through the splines.

Fitting Planks Around Obstructions

Figure 17-15 shows how I use a template to fit planks around a post. With a template I can get a near perfect fit in a difficult spot. Figures 17-15A, B, and C show how to create the template:

1) Cut a piece of cardboard so it fits roughly — within a half inch or so — around the post. An exact fit isn't essential or desired at this point. But be sure the end of the template fits snug against the wall.

2) Cut off the right edge so the template fits snug against the last flooring plank. Next, tack the cardboard in place so it won't move.

3) Use a 1-inch square block of wood to create an offset outline of the entire post. With the cardboard in place around the post, transfer the outline of the post to the template (Figure 17-15A), reproducing the shape of the post at a distance of 1 inch from the post. In this case, I had to place the block and mark the template ten times to recreate the post outline.

4) Now tape the template to the plank intended to fit around the post. Make sure that the right edge and end of the template are flush against the right edge and end of the plank. Remember that the template is offset 1 inch. Use the same 1-inch block to reverse the offset, transferring the lines back onto the plank. See Figure 17-15B.

5) Finally, join the marks with a straight-edge and cut to the line with a slight underbevel. As you can see in Figure 17-15C, the fit will be perfect.

Surfacing the Floor

You can install flooring while other construction work is being completed. But don't bother sanding and finishing until all other trades have pulled off the job. Count on the tile crew grinding spilled grout into the floor; plumbers, electricians and glass setters dropping their tools; and the guy who installs the dishwasher scratching the floor as he slides the appliance into place. You'll be glad you waited.

A Transferring the outline of the post to the cardboard template with a block of wood as a spacer

B Using the wood spacer to retransfer the outline from the template to the flooring

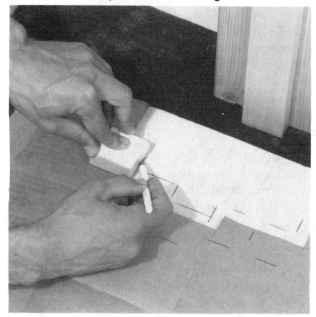

C Is it a perfect fit, or did I cheat and stand the post on top of the uncut board?

Figure 17-15   **Using a template to fit planks around obstructions**

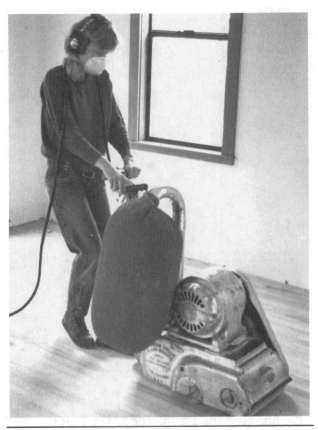

Figure 17-16   **Using a flooring drum sander**

When the strips are laid and all nosings and ending boards are installed, the floor is ready for finishing.

You'll need some specialized tools for this job: a drum and disc sander, a random orbit sander, a floor polisher, and a hand scraper.

Before beginning the process, make sure you do the following:

■ Set any exposed nail heads at least 1/16 inch below the surface of the flooring.

■ Seal off room openings to other areas of the house.

■ Seal off cold air return ducts.

■ Cover counters with plastic tarps and protect the faces of any cabinetry with sheets of cardboard.

■ Thoroughly sweep and vacuum the flooring.

Drum Sanding the Floor

There's only one quick way to get a room of flooring flat and ready to finish — using a flooring drum sander. See Figure 17-16. Beginners think these machines are heavy, noisy, unwieldy dust monsters capable of doing major damage if run out of control. And they're right.

To protect yourself from the noise and dust, use ear plugs and a dust mask. When you start the machine, lean it back so the drum isn't touching the floor. Brace yourself as you lower the drum to the floor. The machine pulls like a tractor.

Here are some tips for using a drum sander:

■ Always keep the machine moving while the drum is in contact with the floor. Tilt the machine back (better machines have a lever that raises the drum) as you approach either end of each sweep. This feathers out the cut.

Remember this: Holding the machine in one place for more than a few seconds while the drum sands away will create a hollow in the floor. Keep the sander moving!

■ Sand toward the wall, then back away from the wall. Note, however, that drum sanders cut faster and deeper when pulled back than when rolling forward. Adjust your technique accordingly. Don't pause at the end of each sweep unless you lift the drum!

■ Most drum sanders are designed so the drum cuts at a slight angle. The cut will be slightly deeper on one side, usually on the left. This lets you feather each pass of the sander as you move up and down the room. If the tilt of the drum is to the left, move the machine from right to left as you make each sweep. Figure 17-17 shows a good sweep sequence. Each sweep should overlap the last by about two-thirds the width of the machine.

■ Always sand with the grain of the flooring. The only exception is when the flooring is radically uneven. In this case, professional sanders make the first sweeps at a 45-degree angle across the grain. But if you're not experienced, I suggest you stay with the grain.

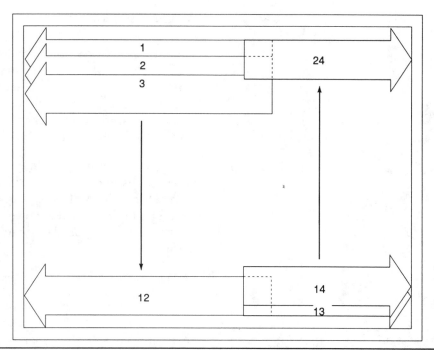

Figure 17-17   Proper sweep sequence using drum sander

Figure 17-18   Disc sander made for floor sanding

Disc Sanding the Perimeter of the Floor

The belt sander can't be used in corners and along the edge of a wall. Here you'll have to use the disc sander. A disc sander made especially for floor sanding works best. See Figure 17-18. I prefer the stability of the larger 7-inch disc over the 5-inch disc. Here are some tips to follow when using a disc sander:

■ Work the machine from left to right in a semicircular motion. Figure 17-19 shows the correct pattern. As with the drum sander, be sure to keep the tool moving while it's running.

■ To feather out the cut at the beginning and end of each stroke, release pressure on the arms. Tilt the machine back on its wheels before beginning the next pass.

■ No matter how good you get with disc and drum sanders, you'll leave some marks between the drum-sanded and disc-sanded areas. To remove these marks, use a random orbit disc sander with 100 grit paper. See Figure 17-20.

Figure 17-19   **Move sander from left to right with semi-circular motion**

Courtesy: Makita, Inc.

Figure 17-20   **Hand-held random orbit sander smoothing out scratch marks between disc-sanded and drum-sanded surfaces**

Figure 17-21   **Using a scraper to smooth out area in hard-to-reach corner**

■ Some areas, such as deep inside corners, are inaccessible to both the disc and drum sanders. Use a floor scraper to surface these areas. See Figure 17-21. Do the final finish with a hand-sanding block or the random orbit sander.

Here's the sanding sequence using both the belt and disc sanders:

1) Start with the drum sander. If the floor is a harder wood, like maple, or if there are noticeable ridges between the strips, start with 24 grit paper on the drum. On other hardwoods, begin with 36 grit. On softwood like pine, begin with 50 grit.

2) Next, use the disc sander around the perimeter of the room where the drum sander can't reach. Use finer paper, usually 60 grit on the disc sander. When finished, vacuum the entire floor.

3) After coarse sanding with drum and disc sanders, apply paste filler to fill nail holes and any gaps between the strips. Instructions for applying filler are in the section that follows.

4) Then fit the drum sander with 60 grit and run over the entire floor. Vacuum the floor.

5) Sand again with 100 grit paper on the disc sander. Vacuum the floor.

6) Finally, switch back to the drum sander fitted with 100 grit paper. When finished, vacuum the floor one last time.

7) Smooth out the transition sanding lines between the disc and drum sander with a random orbit sander.

Applying Filler

After the first pass with the drum and disc sanders, apply filler paste. First vacuum up the dust, then use a trowel to force paste into nail holes and gaps between the strips. Be sure the paste is compatible with the floor finish you plan to use. Instructions that come with the finish material should describe the correct filler material.

If you're using a water-based paste, don't overdo it. Use just enough to fill the gaps and holes. Excess paste sitting on the floor causes wood grain to expand. When the wood dries, you'll find new gaps.

Don't resume sanding until the filler is dry. Check the manufacturer's instructions for waiting time.

Any time you expose a nail head during sanding, reset the nail and apply more putty.

Preparing for Floor Finish

For best results, stain or seal the floor on the same day you do the finish sanding. Wait too long and the flooring grain will probably rise — requiring yet another sanding.

Before applying the stain or sealer, thoroughly vacuum the room. Begin vacuuming with the ceiling, then the walls, and then the floor. Get fine dust out of wood pores by sweeping with a tack rag. Make a tack rag from any clean cotton cloth dampened with mineral spirits and a touch of varnish to make it sticky. If you tie the rag to the end of a broom, your knees will thank you.

Finishing the Floor

Before applying any paste filler, stain or sealer, be sure the materials are compatible. An incompatible finish may never dry. Then you're faced with the nasty task of removing it.

Follow the manufacturer's recommended application procedures. Pay close attention to the suggested drying time between coats. Don't rush it! If you do, the coats won't properly adhere to one another. The finish will eventually crack, craze or crawl — and so will your reputation.

I'll discuss the most common finishes for hardwood floors and the advantages and disadvantages of each. There's no single best floor finish. And no finish is completely impenetrable to all chemicals — or even to water, for that matter.

Penetrating Oils

This is the easiest finish to apply. The oils don't raise the grain very much. They're low sheen, enhance the grain pattern, and give a warm, rich color to most woods. They also create a nice impression of depth.

Unfortunately, oils are not color stable. They darken noticeably in areas that get less light. They don't resist stains very well. For example, foods and household chemicals pass through these oils very easily. A coat of wax adds a little more protection. But keeping wax on any floor requires regular maintenance, and it makes the floor slippery. You can apply shellac or polyurethane over an oil finish, giving a greater depth of color and sense of texture to the finish. The strength, and the durability, of these harder finishes may, however, be somewhat compromised by the underlying oil.

Shellac

Shellac is easy to apply. It dries very quickly and raises the grain only moderately. A three-coat finish gives a beautiful high sheen that's very clear and hard. Shellac is more resistant to staining than oils, though frequent exposure to water turns the finish chalky white. It repairs easily, however. New shellac melts and feathers into the old.

Shellac is also highly color stable. It doesn't darken appreciably with age.

Polyurethanes

There are two types of polyurethane: oil-modified and moisture-cured, and, of course, each has distinct qualities. Both provide a hard, high-sheen surface that's exceptionally resistant to staining. These are also color stable in varying exposure to light. However, they tend to darken with age more than shellac.

Moisture-cured urethanes are harder to apply than oil-modified urethanes. They need a certain level of moisture in the air to go on right. To get good results, pay careful attention to drying times between coats. Also, pay careful attention to the fumes — they're highly toxic.

Oil-modified urethanes are less toxic during the application process than the moisture-cured type. But both are worse than shellac or oils.

The polyurethanes are hard to repair. New coats don't self-feather into the old.

"Swedish Finish"

This term refers to a variety of urea-formaldehyde solutions. Swedish finishes have a high sheen, are extremely hard, and are highly resistant to water, chemical or abrasive damage. But they're not as color stable as the urethanes or shellac.

Also, Swedish finishes are highly toxic during application and for some time afterwards, emitting known carcinogens.

Applying these finishes is a tricky process, requiring careful attention to a variety of factors. Some woods seem to react to the Swedish finish — I've seen fir turn pink under this coating.

Because of environmental restrictions, this finish is not available in many communities.

Waterborne Urethanes

Many improvements were made to these finishes as manufacturers prepared to meet national VOC (volatile organic compound) emission standards. I feel these finishes are as good as, if not better than, almost all of the oil-modified urethanes. A few years ago their resistance to chemical attack was not up to that of the urethanes. But that's no longer the case.

The only drawback I can find is the price — they're probably the most expensive finish you can apply. But considering the overall cost of a floor finishing job, the additional 20 to 30 percent in material costs is nothing compared to the savings in health problems!

Safety note: Always use a filter-type respirator when applying any finish on strip flooring. The type of filter is specified in the Material Safety Data Sheet available from the finish manufacturer or the supplier. It's often printed on the can as well. Although water-based finishes are listed as nontoxic, I recommend wearing a respirator anyway. If the stuff hardens under water (which I've seen it do over time), I imagine it can harden in small particles in your lungs, too. This doesn't seem too healthy to me.

Manhours for Installing Wood Flooring

All figures in the tables are in manhours and are based on the following assumptions:

- Tools and materials needed are available on site.
- The tradesman is a qualified and motivated finish carpenter.
- Work is good quality, stain grade, done no more than 9 feet above floor level.
- All defects are remedied before the carpenter leaves the site.

Add extra time for setup, cleanup, painting or staining, protecting adjacent surfaces, complicated layout or inadequate plans, repair and replacement jobs where fitting and matching is required, working around other trades, setting up scaffolding and ladders for work above 9 feet, and supervision, if necessary. Paint-grade work will usually reduce the time needed by from 20% to 33%.

Wood Flooring

Including layout, vapor barrier and installation with a power nailer. Add for underlayment, sanding and finishing. Borders, feature strips and mitered corners will take more time. Manhours per 100 square feet of floor covered

Tongue and groove strip flooring, 3/4" x 2¼" strips	3.00
Straight edge plank flooring, 3/4" x 6" to 12" wide, set with biscuit joiner plates	5.50
Add for plank flooring set with screws and butterfly plugs	1.00
Filling and sanding a wood floor with drum and disc sanders	1.70
Finishing a wood floor, including filler, shellac, polyurethane or waterborne urethane	1.80

Chapter 18

Stair Trim-out

This chapter describes *some* of what finish carpenters need to know about stairs. Notice that I emphasized the word "some." Stair work is a big topic. A 500-page book couldn't cover every possible stair cutting and stair trim problem. Every stairway can be unique, with its own special challenges and special details.

I'm not going to cover cutting the stair carriage. That's usually done by the framers long before a finish carpenter arrives on site. If you need a book on framing stairways, I can recommend *Stair Builders Handbook*. An order form for this (and other carpentry manuals) is at the back of this book.

What I'm going to cover in this chapter is the basics of stair trim-out — generic information that you'll need on nearly every stair finish job. What you can't glean from this book, you will surely learn from hands-on experience.

Typical Stair Systems and Components

In addition to spiral staircases, there are two other basic types of stairways: open skirt and closed skirt. An open-skirted stairway has at least one side of the stair's treads and risers exposed to view. The side of the stairway adjacent to the wall is snug against a skirt board. Figure 18-1 shows a fine example of this type of stairway. A closed-skirted stairway runs between two walls, hiding the ends of the treads and risers.

High quality treads and risers are usually made of solid hardwood or fir, stained and clear finished. Most utility grade stairs are made from plywood and covered with carpeting.

A "false tread" stairway is a variation of a closed- or open-skirt stairway. It has plywood treads under a carpet runner with finished wood to one or both sides. Again, Figure 18-1 shows an example.

Building codes give us very specific guidance on stairways. That's because stairs are by nature dangerous. Section 3306 of the Uniform Building Code goes into great detail, as do the Southern Standard Building Code and the B.O.C.A. code. The code adopted in your city or county will set standards for tread depth (*run*) and riser height (*rise*). Your inspector will also be interested in distance between the balusters. Some codes may now require three balusters on each tread, rather than the traditional two. Before starting any stair trim-out, be sure the stair carriage and proposed design will meet current code requirements.

Stair trim-out work falls into two categories: the stairs themselves and the balustrade (posts and handrail). Check Figure 18-2 as I describe stair components.

Courtesy: Maizefield Mantels

Figure 18-1   **Open-skirt stairway**

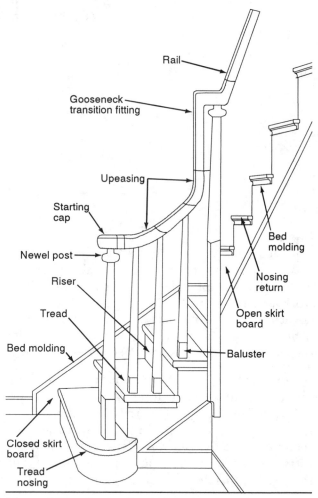

Figure 18-2   **Parts of a stairway**

The Basic Stair Parts

1) Treads should be hardwood unless they're going to be carpeted. Each tread should have a rounded edge along the front called the *nosing*. A tread exposed on one end needs a separate nosing piece on that exposed end. The shape of the nosing should be continuous on all exposed surfaces. A bed molding, such as ¾-inch cove, is usually installed under the tread overhang.

2) Risers, along with the underlying stair carriages, support the treads.

3) A skirt board usually meets the ends of treads and risers on the closed side of the stairway. Treads and risers may be let into dadoes in the skirt board or simply butt against it. On the open side, treads overlay the skirt while the risers are mitered to it. I recommend running a strip of bed molding along the top edge of the skirt board where the skirt meets the wall.

The Parts of the Balustrade

The balustrade consists of the handrail, and the posts and balusters which support it.

There are two basic types of balustrade. One has the rails run between the newel posts, (*post-to-post*). The other type has the rails run over the top of the newels (*over-the-post*), as in Figure 18-2. Over-the-post railing is more difficult to install, but makes a better handrail — especially if you enjoy sliding down balustrades!

The large posts near the beginning, end, and at turns of the railing are called *newel* posts. The smaller posts between newels are called the *balusters*.

The railing may be made of many parts. These components can be grouped into three categories: starting fittings, runs, and transition fittings.

Runs may be straight or curved. In over-the-post construction, the rail begins at a starting fitting which may be a simple cap, or a combination of a cap and easing. A rosette mounted on the wall makes a good end fitting for stair rail.

Transition fittings are used where the rail has to turn (quarter turns, for example), rise (easings and goosenecks), or fit over newel posts (caps). See Figure 18-2.

Preinstallation Procedures

Check the stair carriage before you begin work. Lay a level across the tread cuts on the stair carriage. Are the treads level? Lay the tongue of a framing square against the wall and push the heel (extended with a straightedge if necessary) against each set of risers on the carriage. Are the risers cut at a right angle to the wall?

Then check for adequate spacing between the wall and the carriage on the closed side of the stairs. As a general rule, a 2 x 4 or 2 x 6 placed between the wall and the carriage provides enough clearance for drywall plus a ¾-inch skirt board.

If necessary, adjust the carriages with shims. Fasten these shims with screws so they won't shift under normal load. Then walk up each carriage and listen for any telltale squeaks under your weight. Fix these squeaks by adding more shims and fasteners before they become a permanent part of your stairway.

Sometimes the framers cut stair carriages with treads and risers that are not uniform in height and width. Unfortunately, under nearly all building codes, the difference between the tallest and shortest riser can't be more than ⅜ inch. It's a strict standard, but one that makes sense. A stairway with varying step sizes is dangerous. Because riser heights and tread depths can vary anywhere along the carriage, check each step. Correct the carriage if necessary by recutting or adding to the carriage cutouts.

Don't forget to check the first and last steps. The framer may not have known the thickness of the finish flooring at the top or bottom of the stairway. An unusually thick or thin finish floor could make the first or last step different from all the others. Estimate the thickness of the finish floor covering on the first and last steps. Then see if those risers will be within ⅜ inch of every other riser.

Preparing to Install Newel Posts

To get a secure, wobble-free balustrade, the bases of the newel posts must be secured to something that's rigid. Most of the time they're either bolted directly to the side of a stair carriage or passed through the floor and fastened to a floor joist.

Figure 18-3 shows the newel passing through the first tread, then continuing on through the floor. Notice that the finish carpenter holds the post plumb while marking the outline of the floor cutout. If necessary, modify or move the underlying floor joists so they pass along one side of the cut. This ensures that the newel can be bolted directly to the joist.

When a newel runs alongside a tread instead of passing through it, just bolt the tread to the side of the stair framing. Beef up that area by

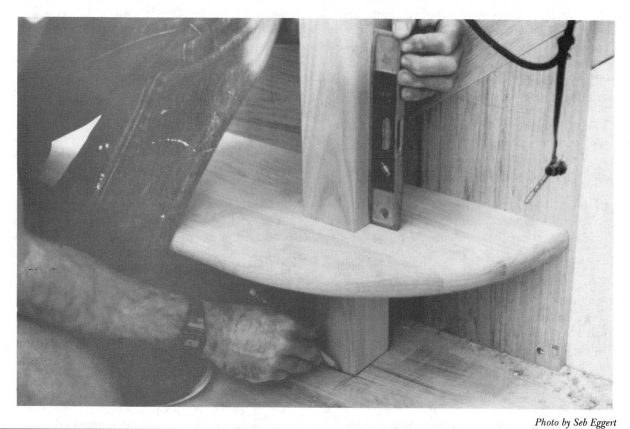

Photo by Seb Eggert

Figure 18-3 **Outline of newel post base marked on floor under the temporarily-installed first tread**

adding cross bracing to a second carriage, as in Figure 18-4. Without this additional bracing, a newel post may eventually twist and loosen the carriage.

Location of newel post

2 x 4 kickboard (removed for clarity on this side)

2 x 6 reinforcement bracing

Figure 18-4 **Reinforcement bracing**

Cutting and Installing Skirt Boards

Begin the job by installing the skirt board to the wall on the closed side of the stairway.

Some stairway specialists prefer to cut dadoes into the skirt board to house ends of treads and risers. There's an advantage to cutting these dadoes: It eliminates unsightly gapping between

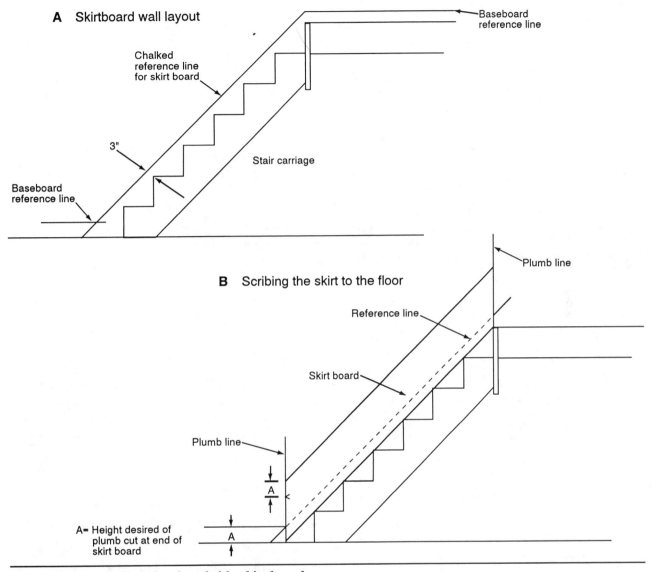

A Skirtboard wall layout

Baseboard reference line

Chalked reference line for skirt board

Stair carriage

3"

Baseboard reference line

Plumb line

B Scribing the skirt to the floor

Reference line

Skirt board

Plumb line

A

A= Height desired of plumb cut at end of skirt board

A

Figure 18-5 **Installing the closed-side skirt board**

the treads, risers and the skirt board if shrinkage should occur. But there's also a disadvantage: This method is time-consuming and requires both special tools and dead accurate layout and cutting.

If the skirt board is thoroughly dry and well secured to the wall, I've found there's rarely a problem with gaps developing there. I simply butt treads and risers against the skirt board on the closed side of the stairway.

Installing the Closed-Side Skirt Board

Installing a skirt board on the closed side of the stairway is a seven-step process:

1) Begin by laying out a profile of the skirt board on the wall. Use chalk to mark a reference line on the wall about 3 inches away from the tips of the carriage step cuts. See Figure 18-5A. The top of the skirt board will be on this line when the board is installed.

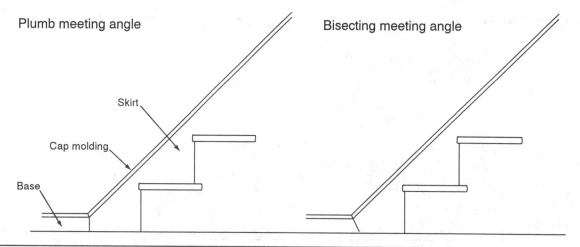

Figure 18-6   **Base and skirt board meeting angles**

2) Next, find end cuts for the skirt board. Mark plumb cut lines on the wall for upper and lower ends of the skirt board. See Figure 18-5B. The meeting angle between baseboard and skirt board is usually plumb. But you may want to cut a bisecting angle as at the right in Figure 18-6.

3) Transfer your cut lines on the wall to the skirt board. Then make the cuts at the upper and lower plumb line.

4) Lay the skirt board against the wall, keeping the plumb cut at either end even with the plumb lines drawn on the wall. Note that the raised board will cover the chalk line that marks its upper edge, as in Figure 18-5B.

5) Set a compass to the height of the baseboard (dimension A in Figure 18-5B) and then this mark dimension on the skirt board. Now draw the floor cut onto the board by holding a leveled straightedge to the dimension mark.

6) Remove the board from the wall and make the level floor cut. Sand out any mill or layout marks with a belt sander. Then use a random orbit sander until the surface is smooth.

7) Now install the skirt to the wall. The top edge should be even with the reference chalk line. Be sure that the end cuts and the face of the board are plumb before nailing it off. If necessary, place shims behind the board at the wall studs to make corrections. Draw the board tight to the wall with nails or screws at each wall.

Installing the Open-Side Skirt Board

Installing a skirt board to the exposed side of the stairway is a little harder. Risers have to be miter-joined into the board. Here's how to lay out cuts for this skirt board.

1) First, mark the position of the bottom edge of the skirt board on the wall with a chalk line. Position the line so that when the skirt board is installed, the top edge comes to a point where the top of the risers meet the underside of the tread. It's a good idea to give yourself another 1/4 inch or so for insurance.

2) Next, draw plumb lines on the wall where newel posts will be installed.

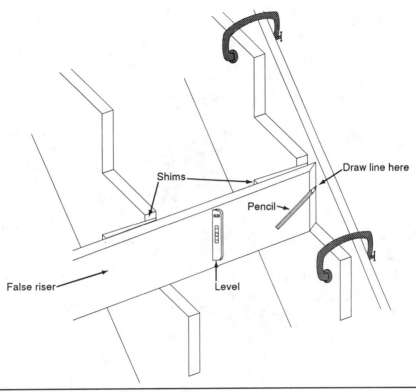

Figure 18-7   **Using a "false riser" to mark cut line on skirt board**

3) Measure the length of the skirt board between newel posts and transfer that length to the bottom edge of the skirt board. Next, use a bevel gauge to transfer the angle of the plumb lines representing the newel posts to the board. Cut the board to length along these lines.

4) Clamp the board temporarily in place and scribe the floor line at the lower end of the board. Remove the skirt and cut to that line. Reposition the board to the chalk line and clamp it in place to the carriage (no fasteners yet!). Be sure the side of the board is plumb. Shim if necessary.

5) Now you're ready to lay out the mitered riser joints. Cut a scrap piece of riser stock just long enough to fit loosely between the installed closed side skirt board and the uncut open skirt. Bevel

one end to make accurate marking easier. I call this board a *false riser*. See Figure 18-7.

6) Place the false riser in position on a step. Make it plumb by setting shims between it and the riser face of the carriage. Then slide it against the inside face of the open skirt board, as in Figure 18-7. Mark a line where the bevel edge meets the skirt. This line is where the inside edge of the miter joint will be cut into the skirt board. Repeat this process on each step.

7) Next, mark a line on the skirt board representing the underside of the tread. Use a false tread similar to the false riser you used earlier. Bevel the marking edge back from the underside face so you can place the pencil on the cut line.

8) Place the false tread in position on the carriage treads. Shim it level side to side and front to back. Now slide it against

Photo by Seb Eggert

Figure 18-8   **Cutting the riser miter joint with a circular saw guided by a jig**

the open skirt will be on the left when ascending, use a worm drive circular saw which tilts to the right. Or you can cut the miter joints with a sharp hand saw. Some finish carpenters prefer to do it that way in the first place. They figure that even when cutting with a circular saw, they'll have to finish off the cuts with a hand saw.

Figure 18-8 shows a finish carpenter cutting the miter joint using a guide clamped to the skirt board. When using a power saw, this guide guarantees a straight line.

Now sand the board clean and reinstall it to the layout marks. Shim the side face of the board plumb, and then nail or screw it in place.

Installing Treads and Risers

Some carpenters prefer to install treads and risers one step at a time, starting at the bottom. Others insist that it's easier and faster to install all the risers first, then go back and set each tread in place. I've discovered that if you install all the risers first, some clown inevitably runs up and down the stairs while you're not looking, mashing the unprotected top edge of the risers!

With the step-at-a-time method, you can lay a piece of cardboard on each completed step as you go up. This protects the stair surface and provides a comfortable perch while you're fitting risers and treads in place.

the outside skirt board. Run a pencil along the bevel edge, marking the cut line. Repeat this procedure on each step.

When marking the outside skirt board is complete, remove it from the wall and lay it across a pair of saw horses, outside face down. If this skirt will be on the right side of the stairs (when ascending), cut the miters with a standard circular saw which tilts angle cuts to the left. If

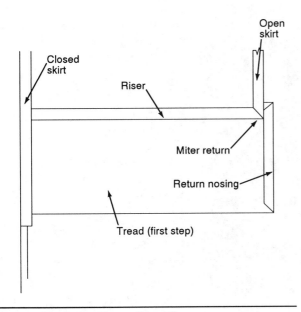

Figure 18-9 **Plan view of tread**

Installing the Risers

If you installed the two skirt boards carefully, keeping side faces plumb and fastening them securely to the walls, then cutting and fitting the risers will be a breeze.

1) Begin by cutting the riser board to rough length. Allow at least 1 inch extra for waste.

2) Cross-cut a 45-degree miter on the end joining the open side skirt. Then measure over from the tip of the open skirt miter to the closed skirt and transfer this measurement to the riser.

3) At the measurement mark, cut a 2- to 3-degree back bevel cross cut in the opposite end meeting the closed skirt. If the skirts are plumb, both cuts can be square to the top edge of the riser. If the open side skirt is off a bit, cut the riser miter square anyway and move the skirt to fit. If the closed side is out of plumb, scribe the riser to fit.

4) Test the fit. Use a sharp block plane to make any final adjustments.

Now it's time to install the riser to the carriages and to the open skirt board. Start by gluing and nailing the miter joint. Then shim the riser level and plumb and nail it off to the carriages. Try to keep the nails within ¾ inch of the edges. That way they'll be hidden behind the tread at the bottom and the bed molding at the top.

Installing the Treads

Treads are a little more complicated to install than risers. Unless the steps are fully carpeted, the exposed end overhanging the open skirt has to be mitered back on itself. Remember that risers above and below each tread must be in place before you can install any tread.

Begin by fitting the closed skirt end. Unfortunately, there's not much chance that the skirt board is perfectly square to the face of the risers. You can assume that you need to scribe the cut. Don't back bevel this cut — it would leave an unsightly tapered gap where the exposed front edge of the tread meets the skirt.

When the tread fits perfectly to the closed-side skirt, mark the bottom of the tread where it overhangs the open skirt. Remove the tread. Lay out the miter joint that caps the front of the return-edge molding. See Figure 18-9. Cut to the line with a circular saw, then finish it off with a hand saw. Test the miter joint fit with a scrap of mitered return-edge molding. If you need to, clean up the cut with a chisel.

You can buy tread stock with the miter return already cut. It's expensive, but it saves time.

For treads on a step with a newel post, lay out and cut a notch where the newel goes through the tread.

Set the tread in place. Level the tread from front to back by shimming the carriages with dry shim stock. Then remove the tread and coat the shims all around with construction adhesive. That helps keep the stairs from squeaking. Reset the tread over the shims and fasten it down with either 8d to 12d finish nails or 2-inch drywall screws. The choice of nails or screws depends on

Photo by Seb Eggert

Figure 18-10   **Miter-returned tread edge molding**

how you plan to countersink or plug the fasteners. Then nail the bottom of the next step's riser to the back of the tread.

Finally, glue and screw the return end molding in place. Make sure you know where the balusters will be before driving these screws. Avoid setting screws where you'll be drilling holes for baluster tenons. Also, for a first class job, the back of the molding should be miter-returned into the skirt board. See Figure 18-10.

In most cases you'll want a bed molding under the tread overhang. Note in Figure 18-10 that I've used a piece of ½-inch by ⅝-inch cove molding. Miter the molding at the outside corner as well as miter-returning the back end — the same treatment you gave to the tread edge molding.

Photo by Seb Eggert

Figure 18-11   **Rail assembly laid out directly on the stairs**

Installing Balustrade

Since a post-to-post handrail is a lot simpler to cut, fit, and install than an over-the-post handrail, I'll cover this installation first.

Installing Post-to-Post Balustrade

Begin the installation of post-to-post balustrade by plumbing the newel posts. Use a long straightedge to make sure the faces of the newel posts are parallel to one another. After securing the posts in place, determine the length and angle cut of the rails in this way:

1) Cut a rail about 1 inch overlength. Lay it on the stairs and slide it against the posts to which it will be attached.

4) Bolt the rail in place at the height specified by your local building code. Generally codes require rail 36 inches above the floor on landings and at least 34 inches above the nose of treads.

Installing Over-the-Post Balustrade

Cutting and fitting over-the-post handrail is a lot more complicated. You have to plan the joint at each post, where rails begin, end, make turns, and gain elevation. Determining the length and angle of these cuts can be a real challenge.

I recommend laying out and dry-fitting the entire assembly directly on the stairs before you install the newel posts. If you didn't mark the location of newel posts when you were installing the treads, now's the time to determine exactly where those posts will go. As a general rule, assume a post is needed at the top and bottom of the stairway, at each turn in the rail, and at 5- to 6-foot intervals along runs of straight railing.

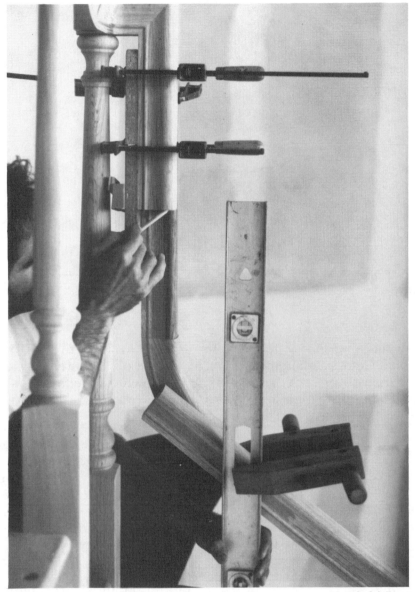

Photo by Seb Eggert

Figure 18-12 **Marking the gooseneck fitting for length with top rail installed**

Figure 18-11 shows a railing clamped directly over the position it will occupy when newel posts are set. Note the unfitted gooseneck riser clamped to the assembly. The gooseneck has to be cut carefully and fitted precisely over the newel post. Otherwise the upper railing won't end up at the right height above the treads. To be safe, you can delay cutting and installing the gooseneck until the upper rail has been installed. See Figure 18-12.

2) Run a pencil along the edge of the post face, marking the cut lines directly on the rail.

3) Cut to the outside of the line with a miter box or chop saw. Test the fit. If necessary, trim the length or angle and prepare the rail to receive the fastening bolt. (I'll describe how a little further on.)

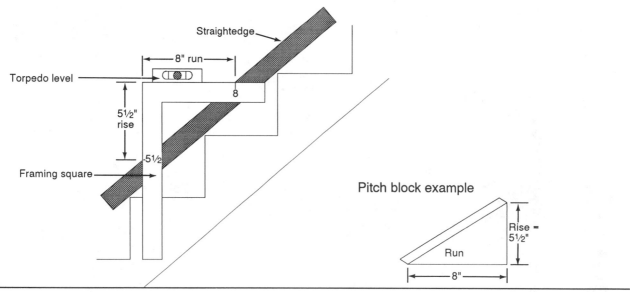

Figure 18-13 **Making a pitch block**

With the gooseneck clamped to the upper rail as shown, lay out the other fittings. Cut the rails to length after cutting the ends of the easings to length and to the correct angle. (I'll describe how below). Position the starting fitting where the first newel post will be installed and clamp the rail to it. Adjust and reclamp until everything is aligned.

Next, mark the rails along the end cuts of the easings. Unclamp the assembly and cut the rails to length with a chop saw.

To lay out the end cuts of up-easings where they join rail that runs up the stairs, I use a pitch block made from a scrap of 2 x 8 (shown in Figure 18-13) to find the correct length and angle.

First, find the pitch of the stairs by laying a straightedge across the nose of the treads. Hold a framing square — leveled with a torpedo level — firmly against it. Read on the square how much the stairs rise over a run of 8 inches. See Figure 18-13. Here, the stairs rise 5½ inches over a run of 8 inches. Use the rise of the stairway as the rise of your pitch block triangle. When the pitch block is laid out correctly, cut it to size.

If the stairway has a landing followed by another run of stairs, check to be sure the pitch of the second rise is the same as the first. If not, make up a second pitch block for this second run of stairs.

To use the pitch block, place the easing fitting on a flat surface. Slide the pitch block, 8-inch base side down, against the underside of the up-easing. See Figure 18-14A. The point where the block touches the easing is the starting point of the cut. Notice that I've marked a cut line on the up-easing at this point.

To find the precise angle of the cut, turn the pitch block so that it rests on the rise side of the triangle, 5½ inches in this case. Then move the block so it's opposite the starting point of the cut line. See Figure 18-14B. Draw a line on the easing along the hypotenuse of the pitch block. This is the angle cut where the rail coming down the slope of the stairs meets the up-easing.

Make the cut on a chop saw if the shape of the fitting allows it to rest on the saw table. Use a block to hold it at the proper angle. (If necessary, refer back to Figure 9-21 in Chapter 9.) If the fitting won't rest in the chop saw, put it in a vise and cut it to length with a sharp crosscut saw.

A   Position of pitch block to find length of up-easing

B   Position of pitch block to find angle of up-easing cut to meet sloped rail

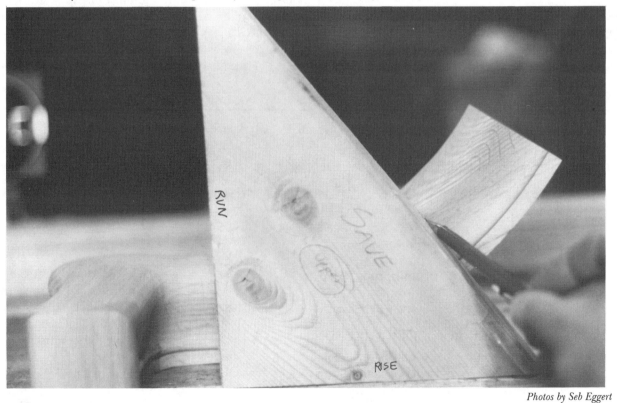

Photos by Seb Eggert

Figure 18-14   **Using the pitch block**

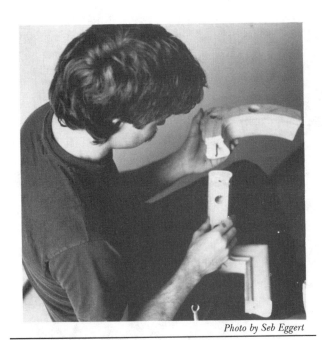

Photo by Seb Eggert

Figure 18-15   Gluing and bolting together the stair parts

Joining Rail Fittings Together

Over-the-post rail fittings are available from most stair parts manufacturers. They're designed to be joined together with 3-inch-long double-ended bolts. One end of the bolt is a lag screw, while the other is a machine screw. The bolt is usually furnished with a star nut designed to be turned by knocking it with a hammer and screwdriver. Because hammering can knock the parts out of alignment, I suggest replacing this nut with a standard hexagonal nut that can be turned with a 12-point box wrench.

Here's how to bolt the fittings together:

1) On the bottom side of both ends to be joined, mark a centerline about 2 inches long. Extend this centerline up the end face of the fittings.

2) Start at the end of the fitting that will receive the lag screw portion of the bolt. This is usually the transition fitting. On the cut face, measure $^{15}/_{16}$ inch up from the bottom of the piece along the centerline. Drill a ¼-inch hole about 2 inches deep. Try to make the hole perpendicular to the end cut face.

3) Screw the lag screw end of the bolt into the hole. This is easier if you lock two hex bolts together on the machine screw end of the bolt. Turn the bolt into its hole with a socket wrench.

4) Drill the end of the opposite fitting (usually the rail) to receive the machine screw end of the bolt:

First, measure back 1⅜ inches along the centerline on the underside of the fitting. Drill a 1-inch diameter hole 1½ inches deep. This is the access hole for tightening the machine screw end of the bolt.

Second, measure up $^{15}/_{16}$ inch on the cut face and drill a ⅜-inch hole until it penetrates the access hole.

5) Use a chisel to make a flat spot along the side of this access hole facing the other fitting. This provides an even surface for the washer and nut to bear against. You can skip this step if the manufacturer provides a special washer with a rounded bearing surface.

6) Slide the parts together and check the alignment. You might have to open up the ⅜-inch hole slightly so the pieces align. A rat-tailed file or a special reaming drill bit is the best tool for this.

7) The fit won't be perfect if the cross section of the rail and the cross section of the transition fittings aren't the same. When you're satisfied that the fit is as close as possible, apply glue to the meeting surfaces. Bolt the pieces together as in Figure 18-15. On a long or complex rail, you may want to join the assembly in sections, especially if it includes a critical fitting such as the gooseneck in Figure 18-12. Finally, plug the nut access hole with a 1-inch wood plug.

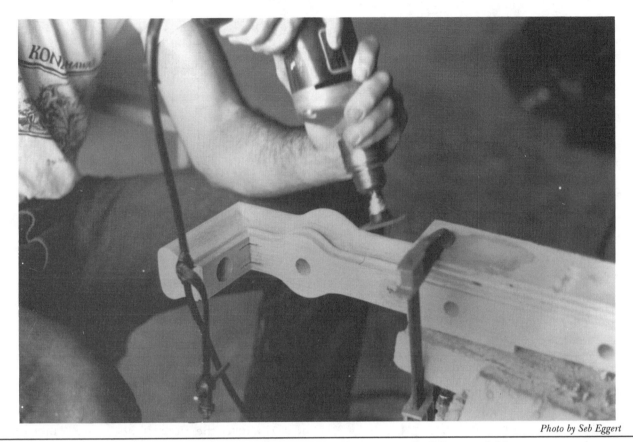

Figure 18-16   **Sanding disc mounted in a drill makes shaping parts together go quickly**

8) When the joint is dry, shape the ends with a rasp, files, and then sandpaper. A small sanding disc mounted on a drill motor (as in Figure 18-16) makes quick work of this task.

Determining Heights of Newel Posts

Figure 18-17 shows the stair rail completely assembled and laid in place on the nose of the stair treads. At this point it's easy to determine the height of the newel posts. Notice in Figure 18-17 that the bottom of the straight rail is just touching the nose of the treads. The starting fitting at the bottom of the stairs and the cap fitting at the landing newel are centered over positions for the newel posts.

The rail to the right of the gooseneck in Figure 18-17 is designed to be 36 inches above the finished floor. The newel height is then 36 inches minus the rail thickness (2½ inches) or 33½ inches.

With the upper rail lifted to 36 inches, the rail height on the stairs is 34 inches above the nose of the treads (see Figure 18-17). Subtracting the rail thickness (2¾ inches along the plumb line) yields a space between the tread nose and the underside of the rail of 31¼ inches.

Notice in Figure 18-17 the 2¼-inch gap under the rail at the balcony. When the handrail assembly is sitting on the tread nosing, the up-easing holds the underside of the balcony railing 2¼ inches above the landing floor. That gap is the difference between the height of the underside rail at the balcony (33½ inches) and

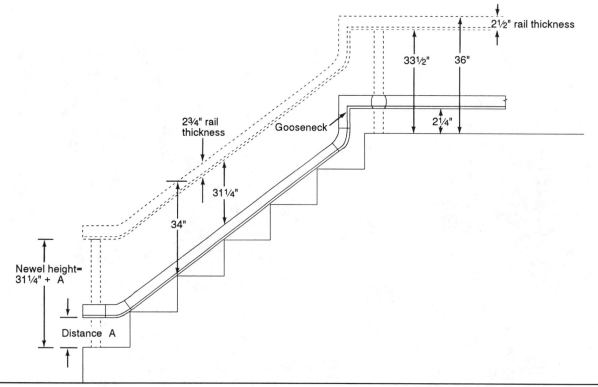

Figure 18-17   Determination of newel post heights

the height of the rail on the stairs (31¼ inches). When the stair railing is lifted 31¼ inches above the nose of the treads, the balcony rail will be 33½ inches above the balcony. Adding the thickness of the rails brings the top of the rail to 34 inches on the stairs and 36 inches on the balcony — just where you want them!

To find the height of the starting newel, measure the distance between the underside of the starting fitting as it lays on the nose of the stairs and the first tread. In Figure 18-17 this is marked Distance A. To this distance add the 31¼ inches we plan to lift the assembly. Distance A plus 31¼ inches is the height of the starting newel between the top of the tread and the underside of the fitting.

Installing the Newel Posts

Cut the newel posts to length as shown above. Be sure to add enough to the shoulder measurement of a notched newel so that it can

be scribed to the floor if necessary. Make the notch deep enough so the centerline of the newel meets the intended centerline of the rail assembly. If the wall isn't plumb, cut the notch at an angle so the newel is plumb when fastened tight to the wall.

Secure a notched newel to the stair carriages with three countersunk 5/16-inch lag bolts. Bolt length depends on the thickness of the wall treatments covering the carriages and the thickness of the notch at the newel post base. The bolt should penetrate the full depth of the carriage. A newel passing through a stair tread (see Figure 18-3) is bolted to the floor joists from below.

While tightening these bolts, monitor the plumb of the post in all directions. When fully driven, plug the countersunk holes on the notched newel with wood plugs. If you're working with stain-grade wood, be sure the plugs match the newels in both color and grain direction.

Figure 18-18 **Jigs for locating baluster mortises on stair treads and rails**

start by setting the assembly temporarily on top of the newel posts. Don't force the top of a newel post out of plumb to make a tenon slip into the mortise hole of a transition fitting. Instead, file the tenon on the side that's missing the hole.

First, lay out baluster holes on the treads. Most building codes require two balusters on each tread, though some may now require three per tread. In any case, the balusters should be spaced evenly on each flight of stairs.

Place the first baluster on each tread so the baluster face on the downstairs side is flush with the face of the riser just below. If that leaves the first baluster too close to the bottom newel post, move the baluster an inch or so back from the nosing on each tread. See Figure 18-1. Then center the other baluster (or balusters) between the first baluster on each tread. When you've marked baluster positions on all treads, use a plumb line to find the correct spot for the top tenon on the underside of the rail.

Figure 18-18 shows a drilling and layout jig for handrails. A jig like this speeds up the work and makes it easier to hold the drill bit at the proper angle without any guesswork. Figure 18-19 shows a close-up view of a jig used to locate holes on the treads. These jigs are available from *By George Enterprises Inc.*, P.O. Box 757, Quakertown, PA 18951 (800-545-9821).

The advantage of these jigs is that they're adjustable and they don't wear out. If you don't want to buy a jig like this, make one out of

Laying Out and Drilling Holes for the Balusters

When the newel posts are installed and the rail is ready to be attached, it's time to begin working on the balusters. We'll begin by laying out holes for the bottom tenons and the top tenons of the balusters.

If the stairway has a post-to-post railing, temporarily bolt the rails to the newel post side faces. If you're installing an over-the-post railing,

Figure 18-19 **Closeup of jig for locating mortise holes on treads**

scrap. Just cut some holes at the proper angle in a scrap of wood and clamp it to the underside of the rail. Replace it when the holes enlarge.

Figure 18-20 shows a sample layout for balusters on a landing or balcony. To find the centerline of the balusters that creates equal spaces between each baluster, and between the outside balusters and the newel post (or wall), follow these steps:

1) Measure the distance between the inside edge of the newel posts (or newel post and wall). In Figure 18-20 it's 30 inches.

2) Add the width of the one baluster. That's 2 inches in the example.

3) Assuming you want center-to-center spacing to be about 4 inches (but check your code), divide the total run by 4 inches. Here, that's 30 inches plus 2 inches divided by 4 inches or 8. Round off to the next highest number if the answer is a fraction. This is the number of spaces between balusters. The number of balusters will be this number less one.

4) Now divide the sum of the inside distance, plus the width of one baluster, by the number of spaces between balusters. The result is the centerline spacing distance between each baluster. In Figure 18-20, 32 inches divided by 8 equals 4 inches.

5) To lay out the centerlines, begin from the inside face of a newel post. The distance between the first baluster centerline and the face of the newel is

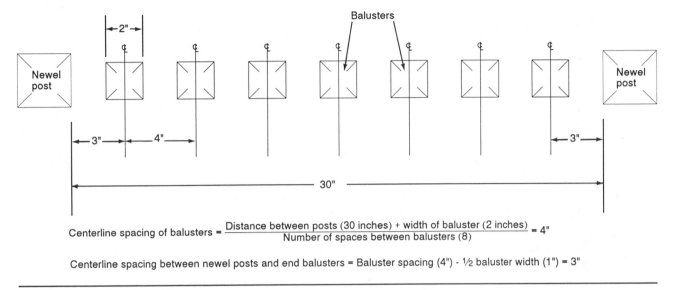

Centerline spacing of balusters = $\dfrac{\text{Distance between posts (30 inches)} + \text{width of baluster (2 inches)}}{\text{Number of spaces between balusters (8)}}$ = 4"

Centerline spacing between newel posts and end balusters = Baluster spacing (4") - ½ baluster width (1") = 3"

Figure 18-20 **Spacing of balusters**

the spacing distance we determined above, less ½ the thickness of the baluster. In the example, 4 inches less 1 inch equals 3 inches. Lay out the remainder of the balusters to the centerline spacing distance.

Installing the Rail Assembly

With all the holes drilled in the treads and landing floors, it's time to cut the balusters to length. The rail assembly should still be bolted or clamped in place temporarily. Measure the baluster length at each location. Then cut and number the baluster for that position. I recommend cutting balusters ⅛ inch or so shy of the measured length. That usually prevents premature "bottoming out" of the railing. If you don't, you could end up with a crowned railing that's impossible to drive down flat.

Once all balusters are cut to length, remove the railing assembly from the newel posts. Now comes the hard part — getting all those balusters seated in their proper homes! I suggest testing

the whole assembly dry before you apply any glue. Go through the following steps, first to last, without any glue. Disassemble the job, make any corrections needed, and do it again with glue. This may add an hour to the job. But it's cheap insurance. You'll spot the errors — such as a bowed or misaligned railing — while it's still easy to make repairs.

1) When you're ready to begin gluing, start with the bottom tenon of each baluster. Find the baluster numbered for each hole in the treads. Put a dab of construction adhesive on the bottom tenon and around the top ¾ inch of the hole. Install the baluster in the hole, spinning it a little in the bottom mortise to spread the glue around. When all the balusters are glued and sitting in the treads, go on to step 2.

2) If this is an over-the-post handrail, spread a dab of glue on each of the newel post tenons.

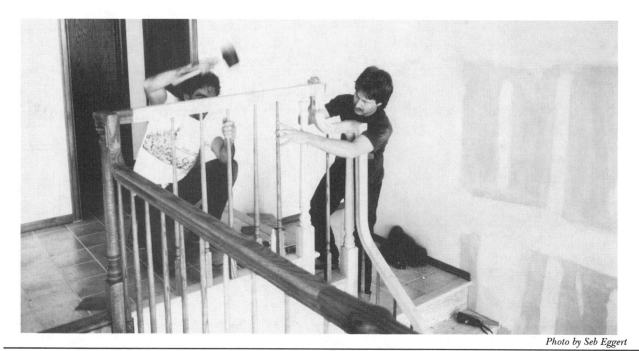

Photo by Seb Eggert

Figure 18-21 **Seating rail assembly into place with a rubber mallet**

3) Get all the help you can muster. (An intelligent octopus friend would be a great asset for this project.) Move quickly to set the rail assembly in place against or over the newel posts and over the balusters.

4) Work your way from one end of the rail run to the other, having your helper slide the balusters part way up out of their bottom mortises and into their holes underneath the rail. Press everything down into place when all the balusters have found their way into holes under the rail.

5) Use a rubber hammer to seat the railing in place over the balusters and newel posts. See Figure 18-21.

6) Sight along the railing to be sure that it's not crowned. This is your last chance before glue sets. Adjust it straight while you can. If necessary, press the rail down hard on the balusters and then toenail the balusters in place until the glue has set.

7) Before the glue sets up, check to be sure all the balusters are aligned. If they have a square base, be sure it's square to the face and edge of the treads. Because the tenon isn't always centered in the base, you might have to rotate the baluster to get it properly aligned on the centerline.

Now take a break — you've earned it! You've just completed one of the most difficult tasks a finish carpenter is asked to handle. You've also just produced one of the finest examples of the art of finish carpentry!

Manhours for Stair Trim-out

All figures in the tables are in manhours and are based on the following assumptions:

- Tools and materials needed are available on site.
- The tradesman is a qualified and motivated finish carpenter.
- Work is good quality, stain grade, done no more than 9 feet above floor level.
- All defects are remedied before the carpenter leaves the site.

Add extra time for setup, cleanup, painting or staining, protecting adjacent surfaces, complicated layout or inadequate plans, repair and replacement jobs where fitting and matching is required, working around other trades, setting up scaffolding and ladders for work above 9 feet, and supervision, if necessary. Paint-grade work will usually reduce the time needed by from 20% to 33%.

Skirt Board for Stairs

Including layout, cutting, fitting, assembly and installation. Based on a 9' floor-to-floor rise where the stair carriage has already been installed by others. Add extra time for recutting, shimming or adjusting the stair carriage.

Cutting and installing a skirt board on the closed side of a stairway.
Manhours per skirt board

Treads and risers butt against skirt board, no dadoes 2.00

Treads and risers set in dadoes 3.75

Cutting and installing a skirt board on the open side of a stairway. Risers miter-jointed into the board. Manhours per skirt board 5.50

Treads and Risers for Stairs

Including layout, cutting, fitting, and installation of treads, risers and typical molding under the tread nosing. Includes miter return on one side of tread.

Manhours per step 2.50

Handrail for Stairs

Including layout, cutting, fitting and installation. Based on milled hardwood rail, posts and fittings.

Setting newel posts. Manhours per post 2.50

Setting post-to-post handrail. Manhours per 10 linear feet of rail 1.50

Setting over-the-post handrail. Manhours per 10 linear feet of rail 2.50

Add for each transition fitting (turn, easing or gooseneck) 1.75

Balusters for Stair Railing

Including layout, cutting, fitting, assembly and installation. Based on mill-made hardwood balusters.

Manhours per baluster with round top section 33

Manhours per baluster with square top 75

Chapter 19

Troubleshooting

Doing finish carpentry is almost as easy as I've tried to make it sound in the last eighteen chapters. But not always. Doors, in particular, can present a number of problems, any one of which is enough to make us wish we were plumbers instead of finish carpenters. Unfortunately, that's only a fraction of the potential for problems in this line of work.

We are sometimes challenged by the need to fit new work to existing work. We see rooms in which we must install trim where the words *level, square* and *plumb* have yet to be invented. We often face situations where no amount of fastening seems able to draw a piece of molding into place.

While we struggle to do the best job we can, we do occasionally slip and dent (or totally mangle) a piece of molding. And if that isn't enough to make our day, we have to deal with tools that seem to go out of adjustment constantly (just because we have a tendency to drop them every other time we use them).

I've included this chapter to help you through these difficult times. After all, if everyone threw in their tool belts and became plumbers, who would buy this book?

Door Problems

One of the most common problems you'll see in door installations is the work of the dreaded gnomes who come out at night and rehang the doors for you. You discover this when you show up in the morning and survey the job you did the day before. You remember the door that you had absolutely perfect the day before: the nice straight gap lines between the door edge and the jamb; the smoothly operating latch mechanism, and the perfectly fitted stops. But it's not that way today. The margin is tapered, the latch is catching on the striker plate, and the stops don't hit the back of the door evenly. The gnomes, whose names are *hinge sag* and *door warp,* have struck again!

After you recite the magical gnome-neutralizing incantations, here's how to go about fixing their mischief.

Correcting Hinge Sag

As a door swings on its jamb hinges, it exerts a lot of pulling force on the top hinge. The heavier and wider the door, the greater the force is going to be. The result can be hinges that spread apart in short order.

As the hinges spread, the door moves away from the upper jamb on the hinge side. That puts a taper in the once-straight margin line. See Figure 19-1. Enough sag and the latch begins to catch on the striker plate.

Here's an old trick that relieves door sag symptoms almost instantly. Simply open the door and insert a nailset or 16d nail between the leaves of the lower hinge. Hold the nail or nailset as close to the hinge barrel as possible. Then close

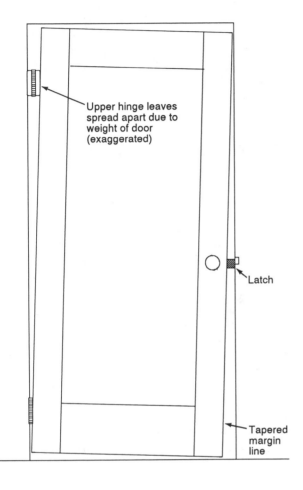

Figure 19-1 **Effects of hinge sag**

the door on the obstruction. That spreads the lower hinge, usually just as much as the upper hinge is spread, eliminating the sag.

Another, perhaps more professional, way to fix hinge sag is to sink one of the upper hinge leaves deeper in its mortise:

1) Measure the amount of taper along the length of the door. Plan to deepen the mortise for the upper hinge leaf by about one-half this distance.

2) Tap out the hinge pins and remove the door from the jamb. Set it in a door buck with the hinges facing up.

3) Remove the upper hinge leaf and deepen the mortise with a router or a hand chisel. If there are three hinges, remove the middle hinge leaf and deepen its mortise as well. Go down only one-half the depth used for the upper hinge.

4) Reinstall the hinges and rehang the door. Make fine adjustments by adding pieces of sandpaper behind the hinge leaves. The finer the paper grit, the smaller the adjustment.

Dealing with Warped Doors

It usually won't happen overnight, but doors can warp in a fairly short time. It's not uncommon for one or two doors on a job to change shape before work is completed. Solid wood frame and panel type doors are most likely to warp quickly.

The cause of warping is usually a door rail or stile with a high moisture content. As the piece dries, it changes shape. The twisting rail or stile often warps the entire door.

When a door warps, the margins at the frame probably won't change. But the door won't close snug against the stops. A warped door will hit the stop along the striker jamb either at the top or the bottom. The other end of the door will be held away from the jamb.

I've found no wholly complete remedy for warped wood doors except replacement. If the warp isn't too bad, plane the stops a little to compensate for the altered shape. If the warp is more severe, remove and reset the stop. Use a pin punch to drive the finish nails through the stop. Reposition the stops against the closed door and re-drive the nails. Be sure to leave a gap the thickness of about a nickel between the stop and the door at the location of the latch. (See Chapter 7.)

If the stops are a part of the jamb and can't be moved, run a rabbet plane along the face of the rabbet. I use a bull-nosed rabbet plane. It makes planing easier at the top of the jamb and at the floor line. See Figure 19-2.

Sometimes a door will warp so much that the latch no longer fits into the hole of the striker plate. Even with the stops adjusted, the latch still may not mesh well with the opening in the plate. In this case, a little filing of the plate opening allows the latch to snap smoothly into place. See Figure 19-3.

A New Door in an Existing Jamb

For those of us who seldom install anything but prehung doors, we may need a little extra coaching. Here are the steps to follow:

1) Check the jamb opening for square and for variations in width. Be sure the new door is larger than the maximum width. And be sure that the door is high enough to take care of any deviation from square.

2) Have a helper hold the oversize door against the jamb. Then mark the outline of the jamb on the door face from the other side. See Figure 19-4.

3) Lay the door across a pair of saw horses or plywood lifts. Use a circular saw to cut to the line.

Figure 19-2   Bull-nose rabbet plane trimming back stop to accommodate a warped door

Figure 19-3   Filing a striker plate opening to accommodate a misaligned latch

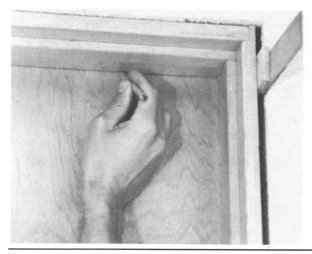

Figure 19-4   Marking the outline of the jamb on the door face

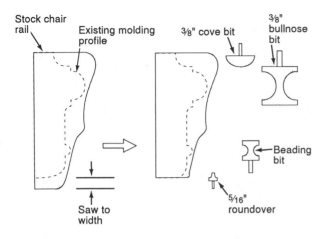

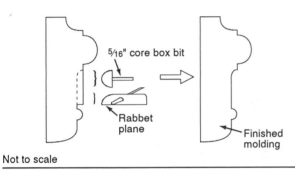

Not to scale

Figure 19-5   **Modification of stock molding profile**

Fitting New Trim to Existing Trim

If you ever do any remodeling work, eventually you'll have to match some existing molding. Often you'll find that the molding pattern you need is no longer sold. That leaves a couple of choices.

The best option is to have molding custom made. Find a millwright who's set up to duplicate a profile. Be sure to run off at least 15 percent more than you plan to use. And don't throw away the blade when you're done.

The second option is to modify stock molding with some combination of router bits, hand planes and scrapers. Figure 19-5 shows some examples. Feather the joint of the new molding to the existing molding with contoured sandpaper holders. See Figure 11-9 in Chapter 11.

4) Test fit the door in the jamb. You want an even margin all around. Use a hand or electric powered jack plane to adjust the size and consistency of the gap.

5) When you're satisfied with the fit, plane a back bevel into the striker side edge of the door, as shown in Chapter 7.

6) Mark the location of the butt mortises by holding the door in place with shims. Mark the position of the existing hinges on the edge of the door. For maximum accuracy, mark with a knife cut rather than a pencil. Cut the mortises and install the hinge leaves.

7) Install the door to the jamb. To mark the location of the latch mechanism, extend a line over from the centerline of the existing striker plate.

Working Out of Level or Plumb

With a room that's way out of level or plumb, I recommend holding off finish work entirely until the structure has been jacked level and true. There's not much you can do to hide trim run out of level and plumb, especially around doors and windows. If you have no choice, the following suggestions may help:

1) When a floor or ceiling isn't level, parallel the existing structure lines. Don't try to keep the trim level. Look at Figure 19-6. Note that the molding has been tapered, thus advertising the problem.

2) When you have to run molding between one surface that's level and plumb and another that's not, use a sequence of moldings to bridge the gap. Taper each molding slightly so you don't have to make the entire correction at once. See Figure 19-7.

Cornice tapered to level, advertises fact ceiling is out.

Cornice allowed to parallel existing out-of-level ceiling, tends to hide problem.

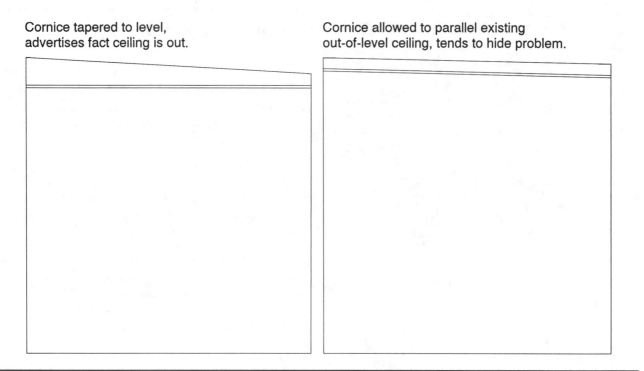

Figure 19-6 **Molding to out-of-level ceiling: two examples**

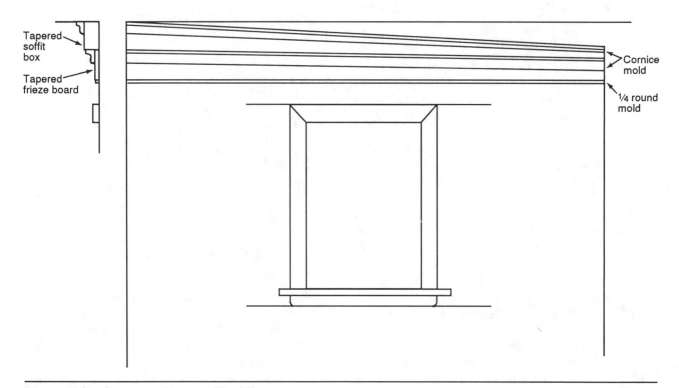

Tapered soffit box

Tapered frieze board

Cornice mold

¼ round mold

Figure 19-7 **Sequenced tapered moldings**

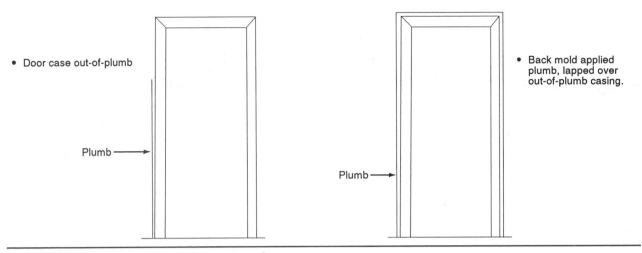

Figure 19-8   **Resolving out-of-plumb door frame**

3) When possible, keep your tapered correction in the molding away from eye level.

4) Look for creative solutions. Figure 19-8 shows an example. In this case, the door frame is far from plumb, and the casing only emphasizes the problem. To hide the problem, apply a little camouflage. At the right, a back molding is rabbeted to overlay the casing, and applied level and plumb.

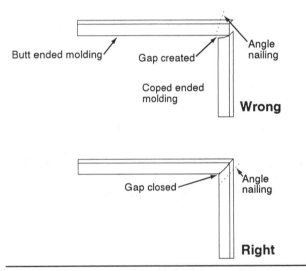

Figure 19-9   **Comparison nailing at coped corners**

Fastening Problems

More often than I care to remember, I've been frustrated by lack of backing for a run of molding. Backing always seems to be missing where it's needed most. For example, backing is critical:

1) In corners where I need to draw moldings together.

2) Anywhere moldings meet at an angle.

3) Where molding ends in the middle of a wall.

Of course, you could always tear off the drywall and install your own backing. Fortunately, there's usually a better way.

When backing is missing at an inside corner, drive a long finish nail at 45 degrees into the corner. The joint here is probably a cope joint. So be sure you're nailing through the molding that's been cope-cut. The nailing angle will actually help draw the joint together. If you goof and nail through the wrong molding, you'll probably open up the joint. Figure 19-9 shows a plan view of nailing molding at a corner.

Figure 19-10   **Bracing molding in place while glue sets**

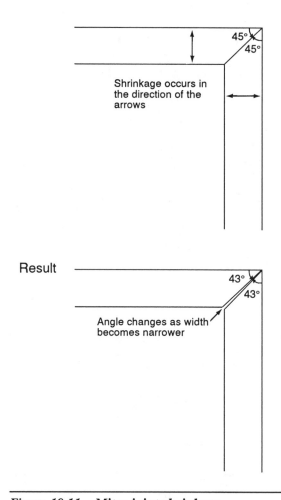

Figure 19-11   **Miter joint shrinkage**

Suppose you're ending a chair rail in the middle of a wall. You probably won't find any backing at the bitter end. It's Murphy's Law that not only won't you find a stud there, but also that the drywall will bow in at that point, making it even worse. One solution here is to use either contact cement or carpenter's glue (aliphatic resin) to glue the molding tight to the wall.

The big advantage of carpenter's glue is that you probably have some on hand already. It's also faster and less messy than contact cement. The drawback is that you need to brace the molding in place while the glue sets. Run a stick from the end of the molding down to the floor. Tack a stop block into the subfloor to hold the brace. See Figure 19-10.

Joint Closure Problems

Sometimes I'll walk onto a site in the morning and notice several joints that have opened up overnight due to wood shrinkage. This rarely happens if you use only properly-seasoned molding for the finish trim. But just in case, here's how to deal with open joints:

1) Miter joints are notorious for opening up with even the slightest wood shrinkage. Figure 19-11 shows an example. In mild cases, driving a screw across the joint should draw it together. With door and window head casings,

Figure 19-12   **Saw recutting a miter joint in place**

to do this. When you refasten the joint, it should close perfectly. If not, repeat the process one more time.

3) Gaps are common at butt joints where a side casing joins a head casing. See Figure 19-13A. Usually the cause is a head casing that's shrunk in width. With luck, the fix is simple. Just drive a finish nail through the head casing down into the side casing to pull it down.

If that doesn't close the gap, you may have to remove and reset the entire head casing. But before doing that, try this trick. Make a thin wedge of wood from the same material as the molding. Be sure the grain along the top of the wedge is similar in direction and appearance to the grain of the head case.

Inject a little glue in the gap and gently tap in the wedge. See Figure 19-13B. Wipe away any excess glue and carefully chisel the wedge flush to the surface of the molding. See Figure 19-13C.

4) As I mentioned earlier, cope joints can open up when they're nailed off to the wall. If angling nails toward the corner through the coped molding doesn't close the joint, try this:

drive the screws down from the top to hide the heads. Otherwise you'll have to countersink and bung the hole.

2) A screw won't do the job if the joint has been glued or if the gap is simply too large. In that case, try this old boat builders' trick. Remove any fastening in the joint. Use a pin punch to poke out any finish nails crossing the joint. Then saw along the line of the joint with a fine-back saw. Figure 19-12 shows how

- Drive shims behind the butt-ended molding until it's drawn tight to the coped molding.

- Use tapered shims to bring cornice molding slightly away from the ceiling or wall. Of course, you now have to deal with the gap created between the molding and the wall surface. See the next step.

5) And then there's always caulk and putty. In paint grade work, fill gaps with beads of caulk and fill nail holes with painter's putty. In stain grade work, caulking is still possible. But the color of the caulk has to match the stain.

The best putty I know of for filling nail holes and small joint gaps is the colored clay type found at most supply yards. These clay mixes have an advantage over hard mixes or putty sticks. You can easily mix colors to get a perfect match. They're also easier to apply — just use your finger. Finally, they don't need sanding, they're long-lasting, and resist cracking and powdering.

Damage Control

I call them *grannies*. Those are the dents my hammer leaves when I miss the nail. Some molding profiles are so fragile that a misplaced hammer blow can cause the molding to literally disintegrate. You can expect these little disasters to happen just as the general contractor or the homeowner shows up on the job. See Section 13 of Murphy's Law.

Here are a few good tricks I've learned for hiding the damage.

Dealing with Hammer Dents

In paint grade work, you can always fill dents with a water-based putty. I use either Durham's *Rock Hard* or a mixture of fine sawdust and glue. For a smoother filling that feathers out better at the edges, try *Poxywood*. This is a new resin-type filler that takes paint just like wood. You can also try resin-based *Bondo* auto-body filler, but it makes the paint look shinier. If you use either product, be sure to wear a respirator and ventilate the room.

A Gap in joint between side case molding in header

B Gap filled with a wedge of wood

C Wedge planed and sanded flush to surface

Figure 19-13 **Repairing a gap in butt joint**

Figure 19-14   **Iron on damp rag raises dented wood fibers**

Making grannies disappear in stain grade work isn't as easy. One solution is to get the crushed wood to expand to fill the depression. Steam applied to wood will raise the wood fiber, sometimes enough to turn the dent into a bump. Here's how to do it:

1) Dampen a cotton rag with water and place the rag over the depression.

2) Press the rag with a hot iron. See Figure 19-14. I use a small iron of the type sold in hobby stores for applying plastic coatings to model airplanes. This iron works great on smaller moldings and doesn't take up much room in my toolbox. Maybe best of all, it's easy to hide from curious eyes.

3) If you don't have an iron, a hammer heated with a torch works almost as well. But don't overheat the head; it can set the rag on fire! And don't use your good hammer. Heating tends to ruin the temper — yours and the hammer's!

4) When the wood is thoroughly dry, sand the surface flush with fine sandpaper on a contoured sanding block. The raised wood is a little softer than the surrounding fibers, so don't sand

without the block. Holding the sandpaper in your fingers tends to hollow out the dent again. Then you're right back where you started.

Dealing with Blow-outs

What if your granny turns the molding to splinters? Suppose there's no wood fiber left to raise with steam? The answer is a *dutchman* — a piece of wood craftily inlaid into the molding. If you cut the piece properly and install and shape it carefully, the dutchman will be a perfect match. Only the practiced eye of another finish carpenter will notice it. You'll have to restrain yourself from showing the dutchman to everyone who walks on the job! Here's how to do a dutchman:

1) Find a piece of scrap wood of the same species as the damaged molding. Try to match the grain as well as the color. Draw a diamond on the scrap large enough to cover the damaged area. See Figure 19-15A.

2) Cut out the diamond to the marked lines. Plane a 2-degree underbevel along the edges. Then lay the dutchman over the damaged area.

3) Use a knife to trace around the perimeter (Figure 19-15B).

4) Use a router fitted with a flat bottom grooving bit to rough out the diamond-shaped recess on the stock. Cut only about 3/16 inch deep. Use a chisel to cut the recess out precisely to the knife lines. See Figure 19-15C.

5) Run glue along the edges of the diamond-shaped piece of scrap. Press it into the recess. Since the lines were drawn along the underbeveled bottom

A Damaged molding with diamond-shaped dutchman drawn on a piece of scrap

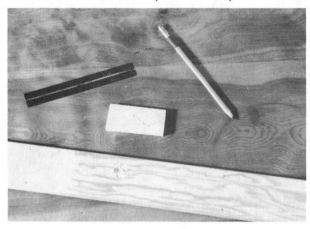

B Dutchman traced onto stock with a scratch-awl

C Diamond-shaped recess cut out of stock

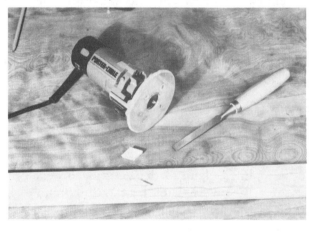

D View of finished repair

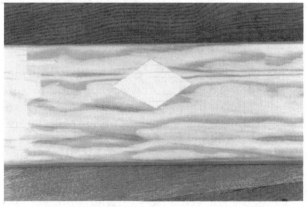

(Contrasting wood used for purpose of illustration)

Figure 19-15　**Installing a dutchman**

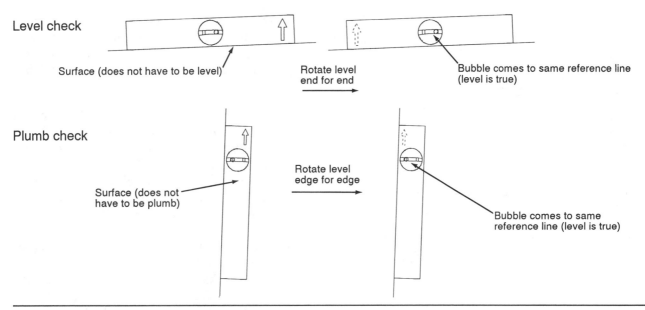

Level check

Surface (does not have to be level)

Rotate level
end for end

Bubble comes to same reference line
(level is true)

Plumb check

Surface (does not
have to be plumb)

Rotate level
edge for edge

Bubble comes to same
reference line (level is true)

Figure 19-16 **Bubble level test procedure**

of the scrap, it'll be slightly oversized. Pressure from a clamp should create a perfect fit all around.

6) When the glue has set, use chisels to carve the dutchman to the shape of the original molding profile. Switch to sandpaper held in a contoured block to smooth and feather the patch so it blends with the molding.

Figure 19-15D shows the finished result. To make it show in the photograph, I've used lighter wood with a different grain for this dutchman. If I hadn't, the dutchman would have been nearly invisible!

Tool Adjustments and Tune-ups

To do professional-quality work, you need professional-quality tools. But even the best tools have to be kept in good condition to get the best results. Unfortunately, there's seldom enough time to repair and calibrate tools when you're in the middle of a major project. That's why I recommend

arriving on the job with tools ready to perform the way they should. A little time invested in getting ready for work can save hours later.

Checking and Adjusting Levels

If you own one of those new electronic levels, be sure to follow the instructions that come with the tool. Calibrating an electronic level isn't difficult. But it takes time, patience, and exact attention to the proper sequence. There's a shorter version of the calibration process explained in this manual. I recommend doing the shorter version on a frequent basis to assure that the tool is reading true.

It's easier to adjust a bubble or spirit level. To determine if it needs adjusting, hold the level against a flat horizontal surface. It doesn't have to be perfectly level. Note the position of the bubble in the vial. Now turn the level end to end (Figure 19-16) and place it in the same position on the surface. Again note the position of the bubble. If the level is in adjustment, the bubble will be in exactly the same position. On a level surface, the bubble should be centered perfectly in the vial.

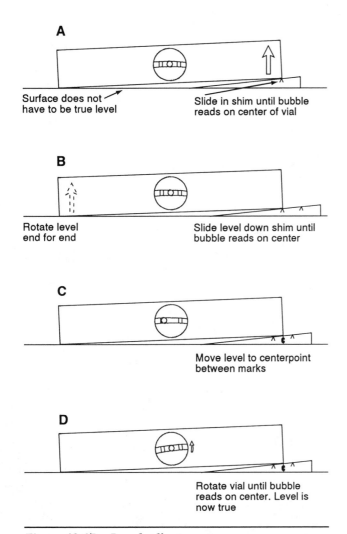

Figure 19-17   **Level adjustment sequence**

Test the plumb bubble the same way. Hold the level against a flat vertical surface. Instead of turning the level end to end, however, rotate it in place 180 degrees. See Figure 19-16. A new edge is now against the surface. Check the position of the bubble.

If the level needs adjustment, here's how to do it:

1) Set the level on a horizontal surface. Slide a long, smooth, evenly-tapered shim under one end of the level until the bubble centers. Mark on the shim the

position of the end of the level. See Figure 19-17A. Also mark the upper edge of the level with a piece of tape so you remember which edge should be up.

2) Turn the level end for end and set the tool to the mark on the shim. Since the bubble is out of adjustment, it won't be in the center. Slide the level up or down the shim until the bubble centers in the vial. Mark where the end of the level falls on the shim. See Figure 19-17B.

3) Remove the level from the shim. Find the point exactly halfway between the two marks. Mark this as the centerpoint. Set the level back on the shim with the end on the center mark. See Figure 19-17C. The level is now setting on true level.

4) Adjust the bubble by loosening the set screws. Rotate the vial until the bubble is exactly centered. See Figure 19-17D. Tighten down the screws. Then check the accuracy of your work by turning the level end to end. When the bubble is centered in the vial, the end should be right on the center mark of the shim.

5) Follow the same steps when adjusting the level for plumb. The only difference is that the test surface and shim are vertical and you'll be rotating the level edge for edge instead of turning it end for end.

Checking and Adjusting Framing Squares

The chances of your framing square being exactly square are a lot less than you might think. I've seen framing squares offered for sale that aren't even close to square. Squares that were accurate when sold probably aren't after they've been dropped or left in the hot sun for several hours. Luckily, there's an easy way to test and adjust a square for absolute accuracy.

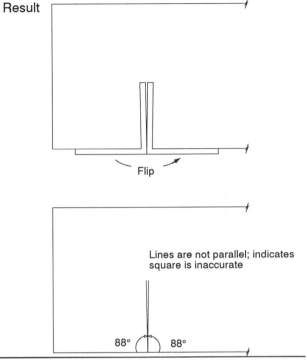

Result

Flip

Lines are not parallel; indicates
square is inaccurate

88°   88°

Figure 19-18   **Checking a framing square**

1) Hold the long leg against the factory edge of a sheet of plywood or MDF. Draw a line along the outside edge of the arm. Now flip the square over. Hold the leg tight to the same edge. Again draw a line along the arm. See Figure 19-18. The two lines should be precisely parallel. If they aren't, note which way you need to move the arm.

2) Correcting it couldn't be simpler. Just hit the thing with a hammer, but not just any hammer and not just anywhere. Use the ball of — you guessed it — a ball peen hammer. And apply hits at the corner.

3) Hold the corner of the square over an anvil. Suppose the tongue needs to move *away* from the leg. Hammer away at the inside corner (Figure 19-19). If the tongue needs to move *toward* the leg, hammer on the outside corner.

Figure 19-19   **Ball peen hammer applied to inside corner of a framing square**

Hammering on the inside or outside corner spreads the metal at that point, moving the tongue and arm apart or together. It doesn't take much hammering to get results either way. Remember, both sides are moving at the same time.

4) After a few strokes, go back to step one to check the progress.

Checking Your Chop Saw for Square

Most chop saws come ready to make accurate crosscuts. They're usually right on the money in the vertical direction. Generally, the stops at the 90, 45, and 22½ degrees are also decently accurate. But drop a chop saw off the tailgate of a pickup and you've got some adjusting to do.

Here's a quick and foolproof way to test a chop saw for square when cutting at either 90 degrees or 45 degrees:

To check for a perfectly vertical cut:

1) Select a 2-foot-long board at least 1½ inches thick. Make a mark along one edge at both ends. Cut the board in half.

2) Lay the two halves on a flat surface. See Figure 19-20. Hold the cut ends together. Flip one of the boards over so the edge mark faces down. If the vertical cut was perfectly square, the board ends will fit together perfectly, without a gap.

To check for horizontal squareness:

1) Select a 2-foot-long board 4 to 6 inches wide. Any thickness will do, but be sure the two edges are parallel. Make marks on both ends to indicate which side is facing up. Lock the chop saw at the 90-degree stop. Then cut the board in half.

2) Set the pieces against a straight edge such as the factory-cut edge of a piece of plywood. Then flip one of the boards over.

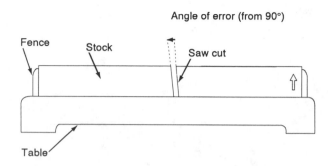

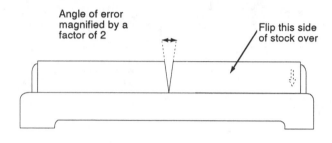

(Error is exaggerated for purpose of illustration)

Figure 19-20   **Checking chop saw for vertical squareness (to table)**

3) If the cut was made at exactly 90 degrees to the back fence of the chop saw, the boards will fit together perfectly. If there's a gap, it will be magnified by a factor of 2. See Figure 19-21. Even a small discrepancy will be obvious.

To check the accuracy of a 45-degree crosscut:

1) Select a 2-foot-long board 2 to 3 inches wide with parallel edges. Lock the chop saw at the 45-degree stop and cut the board in half.

2) Lay the boards on a flat surface. Flip one of the boards over and hold the two mitered faces together. See Figure 19-22. Hold a newly-adjusted framing square against the inside edges of the boards. If the cut is a perfect 45 degrees, the two boards will form a perfect 90-degree angle.

Adjusting the Chop Saw

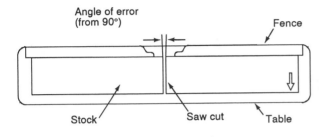

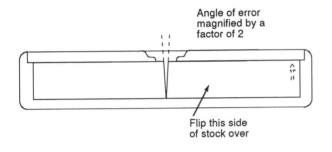

Figure 19-21 **Checking chop saw cut for squareness to fence**

Most chop saws have some way of adjusting the angle between the saw arm and the angle stops. Usually there's a pair of adjusting screws on either side of the base of the arm. Check your saw's manual for instructions. Then check the results using the procedure just described.

There's no easy way to adjust the vertical cutting angle. Accuracy here is determined by machining of the arm base on the pivot point. Luckily, it's not likely to be out of adjustment.

If adjustment is needed, glue strips of sandpaper to one side of the table. That lifts stock to the correct angle. If there's an extension arm, be sure to lift this up as well. Bend it gently in a vise. As an alternative, you can just send the whole business back to the manufacturer for reconditioning.

Figure 19-22 **Checking a chop saw's cut for 45 degrees**

Manhours for Troubleshooting

All figures in the tables are in manhours and are based on the following assumptions:

- Tools and materials needed are available on site.
- The tradesman is a qualified and motivated finish carpenter.
- Work is good quality, stain grade, done no more than 9 feet above floor level.
- All defects are remedied before the carpenter leaves the site.

Add extra time for setup, cleanup, painting or staining, protecting adjacent surfaces, complicated layout or inadequate plans, repair and replacement jobs where fitting and matching is required, working around other trades, setting up scaffolding and ladders for work above 9 feet, and supervision, if necessary. Paint-grade work will usually reduce the time needed by from 20% to 33%.

Correcting a Sagging Door Hinge

Remove door from the jamb, remove hinge leaf, deepen the mortise, rehang the door. Manhours per door	.33

Hang a New Door in an Existing Jamb

Mark a new door to fit the opening, cut the door, set butt hinges, install door in the jamb, install door hardware. Manhours per door	1.75

Chapter 20

Setting Up a Subcontracting Finishwork Business

For more than twenty years I've worked with finish carpenters from one end of the country to the other. In general, I find that we are industrious, conscientious, and highly motivated. But after working several years for wages, many of us decide to go into business for ourselves. Luckily, the same characteristics that make us good tradesmen help make us good subcontractors as well.

I enjoy carpentry subcontracting. I like the freedom inherent in taking sole responsibility for my work. I like knowing that what I create today may last, and be appreciated, well into the next century. Best of all, I like making a good living doing the work I do best.

But be warned, carpentry subcontracting isn't a gravy train. Don't do it to get rich, because you won't. Jumping from being a finish carpenter working for wages to a finish carpentry subcontractor isn't a move to take lightly. To survive, you must learn something of the business of being in business.

Setting Up a Specialty Subcontracting Business

When you contract out finish carpentry, you'll be doing, at least at first, a lot more paperwork than woodwork. You'll have to properly license and insure the business, and pay taxes on business earnings. You'll need to set up

a record keeping system. And you'll need to develop some kind of advertising, at least at first, to put your company name in front of contractors, owners and builders. If you choose to hire employees, you'll face even more paperwork.

Establishing a Legal Identity

There are three forms of business ownership: sole proprietorship, partnership, and corporation. Your business has to be one of the three. Each has advantages and disadvantages.

Sole proprietorship- This is the simplest — you and your business are one. While you are personally liable for everything the company does, you get all the profits. To account for taxes, you'll report income and expenditures on Schedule C of your personal income tax return. Ending or selling the business is a simple matter since there are no partners or stockholders to deal with. But then you alone have to contribute all the money, tools and effort needed to keep the business going.

Partnership- A partnership is a contract between two or more people — an agreement to cooperate toward a common goal. It dissolves when the partners decide that it should end or when a partner dies. Dissolving a partnership can be a complex problem. That's a substantial disadvantage. A second disadvantage is that all partners are liable for 100 percent of the debts — even debts one partner incurs without the knowledge of other partners. The partnership itself pays no taxes. All profits flow through to the partners, who then report their share of partnership earnings on their individual income tax returns.

Of course, there are some solid advantages to having partners. For one, you have someone to share the work load. Each partner can focus on that part of the business they do best. For example, your partner may be better at dealing with general contractors and handling paperwork, while you may prefer to stay in the shop with the smell of fresh-cut wood. Your partner may have plenty of contacts and you may have a world class collection of carpentry tools. These are the signs of a perfect match and an ideal partnership.

If you decide to form a partnership, *put it in writing*. Make it clear. Agree in advance how profits and losses are to be shared. Know from the beginning what will happen when, inevitably, one partner wants out. Seek the help of a lawyer to set up a partnership, no matter how close you and your partner-to-be are.

Incorporation- Becoming a corporation is the third option. It increases the paperwork because corporations have to file tax returns and keep accounting records that comply exactly with IRS requirements. There may be tax advantages to incorporating, but there are also many disadvantages. For a more detailed explanation of the forms of business ownership, order a copy of *How to Succeed With Your Own Construction Business*. It's described in the order form in the back of this book.

Before you decide what kind of business setup you want, discuss the subject with your accountant. If incorporation is your choice, you'll almost certainly need an attorney.

Registration and Licensing

Most states now require a contractor's license if you're bidding jobs rather than working for wages. Some states issue specialty licenses for subcontractors. In other states you'll have to get a general contracting license. Check with your state's contractor's licensing board for the details. They're usually listed in the phone book under the heading for state government offices.

Most cities and counties require every proprietorship, partnership and corporation doing business in that city or county to have a business license. Many state governments also require a business license. Some cities and counties also levy an annual tax on business assets.

Your Business Name

If you're doing business as a proprietorship or partnership, you may need to file and publish a *fictitious name statement* in your city or county. You can always do business under your own name. But if your company name is something other than your own name, you have a DBA (Doing Business As). Filing that DBA (and publishing a notice in a paper approved by the county) is assumed to give notice to everyone of the owner's real name.

Insurance

Many states require licensed contractors to carry a license bond to protect their clients. Your insurance carrier will know what's needed. While you're talking with the insurance carrier, discuss general liability insurance and workers' compensation insurance. Most general contractors require that subcontractors carry liability insurance, and most states require that employers carry workers' comp insurance. You may also want to discuss comprehensive insurance (covering loss or destruction of your tools, for example) and medical insurance (covering medical expenses for you and your family) if you can afford it.

Bookkeeping

No business can stay in business for long without bookkeeping records. If you've worked with a computer, you have an advantage here. Many simple-to-use bookkeeping programs are available for well under $100.

Every business needs accurate and up-to-date accounting records. Failure to keep good business records is the most common reason for failure of small businesses. But you don't have to be an accounting genius to keep adequate records. In fact you don't even have to keep the books at all — you can simply use a bookkeeper. In any case, your business needs a record of *all* financial transactions. Here's why:

■ First, the IRS and your state government require it. If you don't have business records that show where the money came from and where it went, the IRS will estimate what tax you owe and charge you accordingly. And they tend to estimate on the safe side — for them. If you hope to claim business expenses, you'd better have records. Make it a habit to save all receipts.

■ Second, you need to know if the company is making or losing money. The financial history of your company is being written every day as money is earned and spent. The company books help you understand and interpret that history. Cash accumulated in your bank account probably isn't the best index of profit or loss. You can't assume you're making money just because there's money in your checking account. Good record keeping shows you how much money you're making or losing and which jobs are making or losing that money.

■ If the company ever has to borrow money, your loan officer will insist on seeing tax records and financial statements for the last two or three years. No records — no loan.

If you're uncomfortable with bookkeeping, get an accountant to set up a system that works for you. In most cases, all you need is a single-entry cash accounting system. In this simple system, you make a single entry for every transaction, income or expenditure, when it occurs. For a finish carpentry subcontracting business with little inventory, this may be all the bookkeeping system you'll ever need.

If you have employees, use an accountant to set up a payroll accounting system and show you how to draw up and file federal tax forms 941 and 940, and state tax forms. You'll have to apply for an Employer Identification Number (EIN). Figure 20-1 shows the form. Your accountant, or the local post office, can supply you with these forms. For a very reasonable cost, a payroll service will prepare weekly payrolls and file payroll tax forms. See "Payroll Services" in your Yellow Pages.

Checking Account

Open a separate checking account for your business. But shop around. Some banks charge more for a business account. Avoid them if your business is still small. To save time, use a *one write* checking system that produces a check register automatically every time you write a check. A one-write system check ledger is probably available from your bank.

Don't use your business checking account to pay personal bills. The only non-business transaction in the checkbook ledger should be the weekly draw (payroll) you write to yourself. Record this under the non-deductible column. To create a permanent record, write a check for every business transaction. Show all business income in the checkbook ledger, even if you intend to transfer it straight into a savings account. *Each month* balance your checkbook with the bank statement.

Business Forms

You'll need a supply of contract forms, business cards, company letterhead, invoices and receipts. Use an invoice like Figure 20-2 to bill your clients. Provide a receipt when you're paid. You can pick up a book of receipts at any stationery store. I like ones that have a place for me to record the amount still outstanding on the account. This serves as a reminder of the account balance.

Form SS-4
(Rev. April 1991)
Department of the Treasury
Internal Revenue Service

Application for Employer Identification Number

(For use by employers and others. Please read the attached instructions before completing this form.)

EIN

OMB No. 1545-0003
Expires 4-30-94

Please type or print clearly.

1 Name of applicant (True legal name) (See instructions.)

2 Trade name of business, if different from name in line 1

3 Executor, trustee, "care of" name

4a Mailing address (street address) (room, apt., or suite no.)

5a Address of business (See instructions.)

4b City, state, and ZIP code

5b City, state, and ZIP code

6 County and state where principal business is located

7 Name of principal officer, grantor, or general partner (See instructions.) ▶

8a Type of entity (Check only one box.) (See instructions.)
- ☐ Individual SSN _____
- ☐ REMIC
- ☐ Personal service corp.
- ☐ State/local government
- ☐ National guard
- ☐ Other nonprofit organization (specify) _____ If nonprofit organization enter GEN (if applicable) _____
- ☐ Other (specify) ▶ _____
- ☐ Estate
- ☐ Plan administrator SSN _____
- ☐ Other corporation (specify) _____
- ☐ Federal government/military
- ☐ Church or church controlled organization
- ☐ Trust
- ☐ Partnership
- ☐ Farmers' cooperative

8b If a corporation, give name of foreign country (if applicable) or state in the U.S. where incorporated ▶

Foreign country | State

9 Reason for applying (Check only one box.)
- ☐ Started new business
- ☐ Hired employees
- ☐ Created a pension plan (specify type) ▶ _____
- ☐ Banking purpose (specify) ▶ _____
- ☐ Changed type of organization (specify) ▶ _____
- ☐ Purchased going business
- ☐ Created a trust (specify) ▶ _____
- ☐ Other (specify) ▶ _____

10 Date business started or acquired (Mo., day, year) (See instructions.)

11 Enter closing month of accounting year. (See instructions.)

12 First date wages or annuities were paid or will be paid (Mo., day, year). **Note:** *If applicant is a withholding agent, enter date income will first be paid to nonresident alien. (Mo., day, year)* . . . . ▶

13 Enter highest number of employees expected in the next 12 months. **Note:** *If the applicant does not expect to have any employees during the period, enter "0."* . . . . ▶

Nonagricultural	Agricultural	Household

14 Principal activity (See instructions.) ▶

15 Is the principal business activity manufacturing? . . . . ☐ **Yes** ☐ **No**
If "Yes," principal product and raw material used ▶

16 To whom are most of the products or services sold? Please check the appropriate box.
☐ Public (retail) ☐ Other (specify) ▶ ☐ Business (wholesale) ☐ N/A

17a Has the applicant ever applied for an identification number for this or any other business? . . . . ☐ **Yes** ☐ **No**
Note: *If "Yes," please complete lines 17b and 17c.*

17b If you checked the "Yes" box in line 17a, give applicant's true name and trade name, if different than name shown on prior application.

True name ▶ | Trade name ▶

17c Enter approximate date, city, and state where the application was filed and the previous employer identification number if known.

Approximate date when filed (Mo., day, year) | City and state where filed | Previous EIN

Under penalties of perjury, I declare that I have examined this application, and to the best of my knowledge and belief, it is true, correct, and complete | Telephone number (include area code)

Name and title (Please type or print clearly.) ▶

Signature ▶ | Date ▶

Note: *Do not write below this line.* For official use only.

Please leave blank ▶	Geo.	Ind.	Class	Size	Reason for applying

For Paperwork Reduction Act Notice, see attached instructions. | Cat. No. 16055N | Form **SS-4** (Rev. 4-91)

Figure 20-1 **IRS Form SS-4: Application for Employer Identification Number**

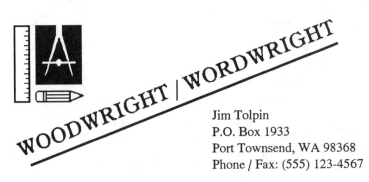

WOODWRIGHT / WORDWRIGHT

Jim Tolpin
P.O. Box 1933
Port Townsend, WA 98368
Phone / Fax: (555) 123-4567

FINE WOODWORKING AND TECHNICAL JOURNALISM

In account with _____ 19____

Terms: Net Cash.   A receipt for bills paid by check sent only upon request.

DATE	REFERENCE	CHARGES	✔	CREDITS	BALANCE
		PREVIOUS BALANCE			

PAY LAST
AMOUNT
SHOWN IN
THIS COLUMN

Figure 20-2 **Sample invoice**

Figure 20-3 **Portfolio presentation case**

Finding Work

Once your paperwork is under control, you're ready to expand the business by venturing out into the marketplace.

If you've been in finish carpentry for several years, you probably won't have to look far. There's a good possibility that contractors you've worked for in the past will want you to bid out your work as a subcontractor. Working as a sub rather than an employee, you eliminate some of their headaches and uncertainties. At least you mean one less person on their payroll.

Extend your search to other contractors who are active in your community. Look in the Yellow Pages under "Contractors-Building" to find likely prospects. Call at least a dozen of these contractors and ask for a few minutes to introduce yourself. Most will be too busy to talk. But you can count on at least a few to make an appointment to see you.

Be sure to take your tools along when calling on contractors on site. In the good old days, a set of properly-maintained hand tools inspected and approved by a foreman was often all a carpenter needed to get a job. Today, a good selection of professional-grade power tools assures the contractor that you at least have the tools, if not the skills, to do the job. To prove the latter, bring along good clear pictures of prior jobs you've done.

Don't make the mistake of thinking a few out-of-focus snapshots stuck in a school folder are going to do it. Instead, put ten to twelve high-quality 8 x 10 color prints of your best work in an attractive portfolio case. See Figure 20-3. These make a much better impression than dozens of second-rate snapshots.

Include in your portfolio case letters of recommendation from satisfied previous employers or homeowners. Hopefully, words like *craftsmanship, reliability, experience, attention to detail, honesty* and *punctuality* will appear in those letters.

Leave a business card with potential clients when you make a sales call. The card doesn't have to be fancy. But it should state clearly what you do and how you can be reached. If you have a contractor's license and are bonded, be sure that information appears somewhere on the card.

Other Job Resources

I've found that the best way to learn about the most active general contractors is to talk with architects. These professionals usually know the builders in your area who specialize in custom homes and better commercial buildings. And it's custom home builders who are more likely to require quality finish carpentry work. If you're going to the trouble and anxiety of making cold sales calls, start with the best and work your way down. You can also get to know your local builders by joining the local building trade association. Once members get to know you and your work, it's unlikely you'll need to knock on unfamiliar doors ever again.

While most of your steady work may come through general contractors, don't ignore the many homeowners who act as their own general contractor. You can often find owner-builders through architects. Another way to find owner-builders is to check the building permit applications at your local planning department. This is public information. These records show the owner's name and address and where, how large and how expensive the project will be. You don't have to be Sherlock Holmes to figure out who needs a finish carpenter.

Advertising

I don't bother advertising with mailers, newspapers, radio, television. I found them a complete waste of time and money. If I haven't found work through the strategies I've recommended, I figure it's time to move. Sometimes it's better to move to where the work is rather than trying to get the work to come to you.

Bids and Contracts

Unfortunately, just finding enough jobs isn't a guarantee that your business will prosper. You've got to learn how to bid the jobs without losing your shirt.

Every finish carpentry job can be broken down into three basic parts:

1) The direct costs of the job itself

2) Overhead costs incurred during the job

3) Profit

Figure the following as direct costs:

- Labor, including withholding taxes and labor-related insurance
- Materials not supplied by the general contractor or homeowner, including supplies such as fasteners, glue, sandpaper, finishing materials and expendable safety gear
- Transportation expenses to and from the job site
- Cost of rental equipment, such as scaffolding or specialized power tools

Figure the following as overhead expense:

- Tools, including their expense, maintenance, and depreciation
- Shop and office rental or mortgage
- Utilities for shop and office, such as heat, lights, phone, and waste disposal
- Vehicle expenses
- Office supplies
- Bookkeeping charges and legal expenses
- Insurance premiums not based on payroll
- License renewal fees and other state, county, and city business fees and taxes

■ Portfolio development and advertising expenses

Determining the Bid Price

Here's the formula for finding the bid price for any type of finish carpentry work:

$$
\begin{array}{r}
\text{Direct Cost} \\
\text{Overhead} \\
\underline{+ \text{ Profit Margin}} \\
\text{Bid Price}
\end{array}
$$

Now let's go through the steps for figuring direct job cost:

1) **Labor Costs:** Estimate the labor hours needed to finish the job. To do this, you'll need to know the amount of materials that must be cut and fit. Prepare a material takeoff from a set of up-to-date plans. Figure 20-4 shows the estimating form I use.

   If you've been in business for a while and have records of installation times, use those figures to estimate installation costs. Otherwise, refer to the sections at the end of each chapter for installation manhours.

   Be sure to add the estimated time to set up, take down and secure materials and equipment each day. On a job with poor site access or poor power access, getting ready to start work can take an hour or more. Be sure to account for this time.

   Multiply the estimated manhours by the hourly labor cost, including insurance, taxes and fringe benefits (such as paid vacations, retirement fund contributions, medical benefits).

2) **Material Costs**: Estimate the quantity of materials and supplies needed for the job. Then price each item. Add the column to find the total.

3) Now add the totals from 1 and 2 above. This is your **direct cost**.

Overhead Costs: Figuring your overhead percentage may be harder. Start with last year's overhead expense, if you were in business last year. Add up all overhead costs. Adjust this total for changes you expect in the coming year. That's your overhead budget for the coming year. If you weren't in business last year, make an informed guess of overhead expense for the coming year.

Divide estimated overhead expense by your expected gross receipts for the coming year. That's your *overhead margin*. For example, suppose receipts are estimated at $200,000 and estimated overhead is $20,000. Your overhead is 10 percent of gross.

Profit: Never think of profit as your wage — that's entirely separate. Profit is what's left over after *all costs and expenses* have been met, including the cost of time you spend both on and off the job.

Most estimators feel that profit margin should vary with the job. Larger jobs usually carry a smaller margin and a smaller job may deserve a larger margin. If you're the only finish carpenter in town and the owner is in a hurry to move in, bump the margin up a little. When work is scarce and you're anxious to stay busy, slice a few percent off the margin.

But I wouldn't cut my profit margin below 5 percent. If I expect to gross $200,000 in the coming year, a 5 percent margin is only $10,000. That's about the smallest return I'm willing to accept on the money I've invested in my company. If, for example, I have $100,000 invested in tools, office equipment, trucks, receivables and work in progress, a $10,000 profit is only a 10 percent return on my invested capital.

Here's an example of how I come up with a bid price by adding overhead and profit margins to the direct cost of a project:

Suppose direct costs are $8,500, overhead is 10 percent and profit margin is 5 percent. Adding overhead and profit percentages together, I get 15 percent (10% + 5% = 15%).

Subcontractor's Estimate Sheet

Date _____   Contractor's License No. _____

Owner's Name _____   General Contractor _____

Address _____   _____

_____   Estimated by: _____

_____   _____

Quantity	Material Description	Material Cost		Labor Operation	Hours	Rate	Labor Cost

Total Material Cost _____   **Total Labor Cost** _____

Sales Tax _____   **Total Materials** _____

Markup _____   **Total Direct Cost** _____

Total Materials _____

Figure 20-4   **Subcontractor's estimate sheet**

I subtract that percentage from 100 percent to get 85 percent (100% - 15% = 85%). Then I divide the direct cost ($8,500) by 85 percent to find the contract price of $10,000.

Here's how the figures come out based on a contract price of $10,000:

Direct cost is 85 percent of the contract price, or $8,500 ($10,000 x 85% = $8,500).

Overhead is 10 percent of the contract price, or $1,000 ($10,000 x 10% = $1,000).

Profit is 5 percent of the contract price, or $500 ($10,000 x 5% = $500).

Direct Cost
Overhead
+ Profit Margin
_____
Bid Price

Direct cost of $8,500.00
Overhead of   1,000.00
Profit of   500.00
_____
Total: $10,000.00

Notice that the procedure I'm recommending may be different from what you've learned elsewhere or what you may have been doing. Many contractors would simply *add* 10 percent for overhead and 5 percent for profit. Let's see how doing that would change our answers and the job with $8,500 in direct costs:

Direct cost of $8,500.00
Add 10% for overhead or   850.00
Add 5% for profit or   425.00
_____
Total: $9,775.00

Note that overhead and profit drop by $225 ($10,000 less $9,775). Using this method, overhead isn't 10 percent and profit isn't 5 percent any more. Overhead ($850) is only 8.7 percent of the contract price ($9,775) and profit is only 4.3 percent of the contract price.

Now test your understanding of the procedure I suggest. Use the figures that follow to find the bid price in dollars and the overhead and profit in dollars:

- Estimated labor cost for installation is $13,250.00
- Estimated labor cost for setup and takedown is $1,750.00
- Estimated cost of materials and supplies is $5,000.00
- Overhead margin will be 10 percent
- Profit margin will be 5 percent

Figure the bid price:

Bid price in dollars   _____
Overhead in dollars   _____
Profit in dollars   _____

When you've filled in the blanks above, check your answers against my answers.

Here's how you should have handled the problem:

1) Remember the bid price formula:

Direct Cost
Overhead
+ Profit Margin
_____
Bid Price

If overhead is 10 percent and profit is 5 percent, direct cost must be 85 percent (100% - 15% = 85%).

2) We know direct cost is $20,000 ($13,250 + $1,750 + $5,000 = $20,000). Dividing $20,000 by 85 percent, we get $23,529.41. That's the bid price.

3) Overhead must be 10 percent of the bid price, or $2,352.94 (10% x $23,529.41 = $2,352.94).

4) Profit must be 5 percent of the bid price or $1,176.47 (5% x $23,529.41 = $1,176.47).

Then, check your answers:

Direct cost of $20,000.00
Overhead of   2,352.94
+ Profit of   1,176.47
_____
Total: $23,529.41

Working for Time and Materials

If you're new to the game, you may not feel very secure about bidding jobs at first. What you leave out of a bid — errors of judgment in labor time, and not accounting for any damaged or miscut materials — can potentially cost you plenty. Even experienced finish carpenters occasionally make mistakes like these. But as you develop more experience, the good jobs and the bad jobs should balance out, leaving you with a fair profit when all work is considered.

While you're new at the game, you may want to sell your work on a *cost-plus* or time and materials basis if you can. That means charging clients for your time and the materials you use plus a percentage to cover overhead and profit. A client who trusts you and knows your work may be happy to do business on this basis — especially a client who expects to save some money in the process! A sharp client will probably see other advantages:

- They can make unlimited changes in your work without writing change orders or getting new estimates.

- They can control quality — you're happy to fix anything that's not right. You'll even tear it all out and start over, if that's what's needed. It's your client's money and they make the decisions.

If you agree to work on a cost-plus basis, get a written agreement. State clearly what your terms are. Show overhead and profit as a percent of the total project cost and show how it's going to be figured. Will it be a percentage added to materials only? Or will it be a percentage of the total contract price, as I recommended earlier? State when invoices will be presented and when they should be paid.

For your own protection, add overhead and profit on each invoice, not just the last invoice. If possible, avoid any "ball park" estimate or not-to-exceed price. These maximums tend to become the contract price, defeating the purpose of working on a cost-plus or time and materials basis.

One final note: Under a cost-plus or time and materials contract, you're still a contractor. You still have to carry all the same insurance, have to be licensed and have the same tax obligations.

The Contract Forms

If you expect to get paid according to your agreement, get the agreement in writing and get it signed. Otherwise you may not get paid the full amount due. Figure 20-5 shows a typical proposal and contract form. Note that state requirements vary. Contract forms sold at larger stationery stores in your state will probably comply with state law. If you're in doubt, get the advice of a more experienced contractor or an attorney. In any case, always describe exactly what you plan to furnish and install. The contract should refer to the plans and specifications if you bid from plans and specs. Don't leave any doubt about what you plan to do. Try to anticipate problems and possible disputes. Add extra pages of description to the contract, if necessary.

Change Orders

Changes are very common in the construction industry. Every owner and nearly every designer is going to think of something they want to change between the time the contract is signed and the work is completed.

Proposal and Contract

For Residential Building Construction and Alteration

Date _____ 19 ___

To _____

Dear Sir:

We propose to furnish all material and perform all labor necessary to complete the following:

_____

_____

_____

_____

_____

Job Location:

All of the above work to be completed in a substantial and workmanlike manner according to the drawings, job specification, and terms and conditions on the back of this form for the sum of

Dollars ($_____)

Payments to be made as the work progresses as follows: _____

_____

_____

the entire amount of the contract to be paid within _____ days after substantial completion and acceptance by the owner, The price quoted is for immediate acceptance only. Delay in acceptance will require a verification of prevailing labor and material costs. This offer becomes a contract upon acceptance by contractor but shall be null and void if not executed within 5 days from the date above.

By _____

"YOU THE BUYER, MAY CANCEL THIS TRANSACTION AT ANY TIME PRIOR TO MIDNIGHT OF THE THIRD BUSINESS DAY AFTER THE DATE OF THIS TRANSACTION, SEE THE ATTACHED NOTICE OF CANCELLATION FORM FOR AN EXPLANATION OF THIS RIGHT."

You are hereby authorized to furnish all materials and labor required to complete the work according to the drawing, job specifications, and terms and conditions on the back of this proposal, for which we agree to pay the amounts itemized above.

Owner _____ **Date** _____

Owner _____ **Date** _____

Accepted by Contractor _____ **Date** _____

Figure 20-5 **Proposal and contract (front)**

Terms and Conditions of Contract

1. The Contractor agrees to commence work hereunder within ten (10) days after the last to occur ot the following, (1) receipt of written notice from the Lien Holder, if any, to the effect that all documents required to be recorded prior to the commencement of construction have been properly recorded; (2) the building site has been properly prepared for construction by the Owner, and (3) a building permit has been issued. Contractor agrees to prosecute work thereafter to completion, and to complete the work within a reasonable time, subject to such delays as are permissable under this contract. If no first Lien Holder exists, all references to Lien Holder are to be disregarded.

2. Contractor shall pay all valid bills and charge for material and labor arising out of the construction of the structure and will hold Owner of the property free and harmless against all liens and claims of lien for labor and material against the property.

3. No payment under this contract shall be construed as an acceptance of any work done up to the time of such payment, except as to such items as are plainly evident to anyone not experienced in construction work, but the entire work is to be subject to the inspection and approval of the inspector for the Public Authority at the time when it shall be claimed by the Contractor that the work has been completed. At the completion of the work, acceptance by the Public Authority shall entitle Contractor to receive all progress payments according to the schedule set forth.

4. The drawings and specifications are intended to supplement each other, so that any works exhibited in either and not mentioned in the other are to be executed the same as if they were mentioned and set forth in both. In the event that any conflict exists between any estimate of costs of construction and the terms of this Contract, this Contract shall be controlling. The Contractor may substitute materials that are equal in quality to those specified if the Contractor deems it advisable to do so.

5. Owner agrees to pay Contractor its normal selling price for all additions, alterations or deviations. No additional work shall be done without the prior written authorization of Owner. Any such authorization shall be on a change-order form, approved by both parties, which shall become a part of this Contract. Where such additional work is added to this Contract, it is agreed that all terms and conditions of this Contract shall apply equally to such additional work. Any change in specifications or construction necessary to conform to existing or future building codes, zoning laws, or regulations of inspecting Public Authorities shall be considered additional work to be paid for by Owner as additional work. If the quantity of materials required under this Contract are so altered as to create a hardship on the Contractor, the owner shall be obligated to reimburse Contractor for additional expenses incurred. It is understood and agreed that if Contractor finds that extra concrete is required he is authorized by the Owner to pour the amount of concrete that is required by the building code or site conditions and shall promptly notify Owner of such extra concrete. Owner shall promptly deposit the cost of the required extra concrete with the Contractor. Any changes made under this Contract will not affect the validity of this document.

6. The Contractor shall not be responsible for any damage occasioned by the Owner or Owner's agent, Acts of God, earthquake, or other causes beyond the control of Contractor, unless otherwise herein provided or unless he is obligated by the terms hereof to provide insurance against such hazards. Contractor shall not be liable for damages or defects resulting from work done by subcontractors. In the event Owner authorizes access through adjacent properties for Contractor's use during construction Owner is required to obtain permission from the owner(s) of the adjacent properties for such. Owner agrees to be responsible and to hold Contractor harmless and accept any risks resulting from access through adjacent properties.

7. The time during which the Contractor is delayed in his work by (a) the acts of Owner or his agents or employees or those claiming under agreement with or grant from Owner, including any notice to the Lien Holder to withhold progress payments, or (b) any acts or delays occasioned by the Lien Holder, or by (c) the Acts of God which Contractor could not have reasonably forseen and provided against, or by (d) stormy or inclement weather which necessarily delays the work, or by (e) any strikes, boycotts or like obstructive actions by employees or labor organizations and which are beyond the control of Contractor and which he cannot reasonably overcome, or by (f) extra work requested by the Owner, or by (g) failure of Owner to promptly pay for any extra work as authorized, shall be added to the time for completion by a fair and reasonable allowance. Should work be stopped for more than 30 days by any or all of (a) through (g) above, the Contractor may terminate this Contract and collect for all work completed plus a reasonable profit.

8. Contractor shall at his own expense carry all workers' compensation insurance and public liability insurance necessary for the full protection of Contractor and Owner during the progress of the work. Certificates of such insurance shall be filed with Owner and with said Lien Holder if Owner and Lien Holder so required. Owner agrees to procure at his own expense, prior to the commencement of any work, fire insurance with Course of Construction, All Physical Loss and Vandalism and Malicious Mischief clauses attached in a sum equal to the total cost of the improvements. Such insurance shall be written to protect the Owner and Contractor, and Lien Holder, as their interests may appear. Should Owner fail to do so, Contractor may procure such insurance as agent for Owner, but is not required to do so, and Owner agrees on demand to reimburse Contractor in cash for the cost thereof.

9. Where materials are to be matched, Contractor shall make every reasonable effort to do so using standard materials, but does not guarantee a perfect match.

10. Owner agrees to sign and file for record within five days after the completion and acceptance of work a notice of completion. Contractor agrees upon receipt of final payment to release the property from any and all claims that may have accrued by reason of the construction. If the Contractor faithfully performs the obligations of this part to be performed, he shall have the right to refuse to permit occupancy or use of the structure by the Owner or anyone claiming through the Owner until Contractor has received the payment due at completion of construction.

11. Any controversy or claim arising out of or relating to this contract, shall be settled by arbitration in accordance with the Rules of the American Arbitration Association, and judgment upon the award rendered by the Arbitrator(s) may be entered in any Court having jurisdiction.

12. Should either party hereto bring suit in court to enforce the terms of this agreement, any judgment awarded shall include court costs and reasonable attorney's fees to the successful party plus interest at the legal rate.

13. Unless otherwise specified, the contract price is based upon Owner's representation that site is level and cleared and is not filled ground or hard rock and that there are no conditions preventing Contractor from proceeding with usual construction procedures and that all existing electrical and plumbing facilities are capable of carrying the extra load caused by the work to be performed by Contractor. Any electrical meter charges required by Public Authorities or utility companies are not included in the price of this Contract, unless included in specifications. If existing conditions are not as represented thereby necessitating additional excavation, blasting, plumbing, electrical, curbing, concrete or other work, the same shall be paid for by Owner as additional work.

14. The Owner is solely responsible for providing Contractor prior to the commencing of construction with such water, electricity and refuse removal service at the job site as may be required by Contractor to effect the construction of the improvement covered by this Contract. Owner shall provide a toilet during the course of construction when required by law.

15. The Contractor shall not be responsible for damage to existing walks, curbs, driveways, cesspools, septic tanks, sewer lines, water or gas lines, arches, shrubs, lawn, trees, clotheslines, telephone and electric lines, etc., by the Contractor, sub-contractor, or supplier incurred in the performance of work or in the delivery of materials for the job. Owner hereby warrants and represents that he shall be solely responsible for the condition of the building site with respect to finish grading, moisture, drainage, alkali content, soil slippage and sinking or any other site condition that may exist over which the Contractor has no control and subsequently results in damage to the building.

16. The Owner is solely responsible for the location of all lot lines and shall identify all corner posts of his lot for the Contractor. If any doubt exists as to the location of such lot lines, the Owner shall at his own cost, order and pay for a survey. If the Owner shall wrongly identify the location of the lot lines of the property, any changes required by the Contractor shall be at Owner's expense. This cost shall be paid by Owner to Contractor in cash prior to continuation of work.

17. Contractor has the right to sub-contract any part, or all, of the work herein agreed to be performed.

18. Owner agrees to install and connect at owner's cost, such utilities and make such improvements in addition to work covered by this contract as may be required by Lien Holder or Public Authority prior to completion of work of Contractor.

19. Contractor shall not be responsible for any damages occasioned by plumbing leaks unless water service is connected to the plumbing facilities prior to the time of rough inspection.

20. The Owner is solely responsible for all charges incurred for grading of lot for level building site, removing all trees, debris, and other obstructions prior to start of construction.

21. Owner hereby grants to Contractor the right to display signs and advertise at the building site.

22. Contractor shall have the right to stop work and keep the job idle if payments are not made to him when due. If any payments are not made to Contractor when due, Owner shall pay to Contractor an additional charge of 10% of the amount of such payment. If the work shall be stopped by the Owner for a period of sixty days, then the Contractor may, at Contractor's option, upon five days written notice, demand and receive payment for all work executed and materials ordered or supplied and any other loss sustained, including a profit of 10% of the contract price. In the event of work stoppage for any reason, Owner shall provide for protection of, and be responsible for any damage, warpage, racking, or loss of material on the premises.

23. Within ten days after execution of this Contract, Contractor shall have the right to cancel this Contract should he determine that there is any uncertainty that all payments due under this Contract will be made when due or that an error has been made in computing the cost of completing the work.

24. This agreement constitutes the entire contract and the parties are not bound by oral expression or representation by any party or agent of either party. All changes to the work approved by Owner and accepted by Contractor become part of this contract.

25. The price quoted for completion of the structure is subject to change to the extent of any difference in cost of labor and materials as of this date and the actual cost to Contractor at the time materials are purchased and work is done.

26. The Contractor is not responsible for labor or materials furnished by Owner or anyone working under the direction of the Owner and any loss or additional work that results therefrom shall be the responsibility of the Owner.

27. No action arising from or related to the contract, or the performance thereof, shall be commenced by either party against the other more than two years after the completion or cessation of work under this contract. This limitation applies to all actions of any character, whether at law or in equity, and whether sounding in contract, tort, or otherwise. This limitation shall not be extended by any negligent misrepresentation or unintentional concealment, but shall be extended as provided by law for willful fraud, concealment, or misrepresentation.

28. All taxes and special assessments levied against the property shall be paid by the Owner.

29. Contractor agrees to complete the work in a substantial and workmanlike manner but is not responsible for failures or defects that result from work done by others prior, at the time of or subsequent to work done under this agreement, failure to keep guttes, downspouts and valleys reasonably clear of leaves or obstructions, failure of the Owner to authorize Contractor to undertake needed repairs or replacement of fascia, vents, defective or deteriorated roofing or roofing felt, trim, sheathing, rafters, structural members, siding, masonry, caulking, metal edging, or flashing of any type, or any act of negligence or misuse by the Owner or any other party.

30. Contractor makes no warranty, express or implied (including warranty of fitness for purpose and merchantability). Any warranty or limited warranty shall be as provided by the manufacturer of the products and materials used in construction.

31. The owner shall assume full responsibility for the project and the site beginning on the date the Certificate of Substantial Completion is issued.

32. Any conflict between plans, specifications, and this contract, this contract shall prevail.

Figure 20-5 cont. **Proposal and contract (back)**

Just be sure that changes aren't made at your expense. If the owner walks on the job and asks you to run baseboard in a room where the plans don't show it, no problem! You'll be glad to do the work — at their expense! That extra base is a change to the contract, for which you must be paid. And you *will* get paid if you require a change order. But without one, you're taking your chances. Some owners seem to think a little extra added to the job will be done at no extra charge.

I keep a stash of change orders in my truck and use them whenever I'm asked to do something that wasn't in the original agreement. Since a change order is a new contract, it has to be signed by the same people who signed the original agreement. Figure 20-6 shows the form I use. Remember, unless you're working under a cost-plus contract, get a signed change order for all changes from the original terms of the contract. Also remember to add overhead and profit margins to the direct cost of the change. While changes usually involve relatively small amounts of material, they take a disproportionate share of administrative time. So you're entitled to a higher markup on change orders. If your markup is usually 10 percent of direct costs, markup on change orders should be at least 15 percent. On a small change order, 20 or even 25 percent may not be too much.

Hiring, Managing and Holding On to a Good Crew

There seems to be no easy way to find and keep good carpenters. The better ones are busy at their trade, enjoying plenty of job opportunities. Many of the best won't work for wages. They become subcontractors themselves, earning the extra profit a subcontractor can demand.

Still, I've found some highly skilled workers who don't want anything to do with running a business. They enjoy working on the site, not shuffling papers. They like that regular paycheck at the end of every week. These are the kind of people I like to get on my payroll.

Generally, the best way to find good finish carpenters is to talk to people in the trade. When you see a finish carpentry crew working on a job, stop to chat a while. Ask the crew if they know of any good finish carpenters who are looking for work. Put out the word that you're hiring. Advertise only as a last resort — you'll have to wade through an awful lot of chaff to find even a single kernel of wheat.

Here are some subjects to ask when interviewing potential employees:

- Find out about their experience — how long they've been in the trade.

- Ask about the tools they own.

- Find out why they want to work for you — and why they're looking for another job.

- Ask what their professional goals are.

- Ask what kind of finish work they like to do most.

- Ask what they like and dislike about the trade.

- Ask for references. Follow up with phone calls. Explain to each reference what you expect from an employee. Then ask if the applicant is up to it. Ask the reference if they would rehire this applicant.

Listen carefully to how they answer your questions. It doesn't take a psychic to make educated guesses about how an applicant will perform on the job.

I avoid applicants with drinking or drug problems, though I do consider hiring them after they've licked the problem for at least six months. Today, even small communities have drug testing centers that can perform tests for most drugs and alcohol at modest cost. Remember that you're in the carpentry business, not the rehabilitation business. Be aware that anyone who works for you represents your company. Don't gamble away jobs with flaky help!

Change Order

Change order # _____

Date submitted: _____

Project name: _____

Date of contract: _____

Project location: _____

Owner: _____

Architect: _____

Contractor: _____

This change order authorizes the Contractor to make the following changes in the work:

Work Changes	**Contract Adjustments**		**Contract Time** (days)	
description or reference document(s)	$ credits	$ deductions	extension	reduction

Totals $ _____ $ _____

Net Totals   +/— $ _____ . _____   +/— _____ **days**

Amount of original contract:	$ _____ . _____	Time to complete original contract:	_____ days
Amount of contract before this change order:	$ _____ . _____	Time to complete before this change order:	_____ days
Net total this change order:	+/— $ _____ . _____	Net total this change order:	+/— _____ days
Net contract amount:	$ _____ . _____	New contract time:	_____ days
		Time since commencement:	— _____ days
		Time remaining:	_____ days

Approved by: _____

                                                (owner)   (date)

                                      (owner's signature)

Approved by: _____

                                                (contractor)   (date)

                                      (contractor's signature)

Figure 20-6   **Change order**

Holding On to the Good Ones

Once you have good employees on board, do your best to keep them. Show appreciation for the work your people do and the way they conduct themselves. The best form of appreciation, of course, is to pay them well. If possible, pay a little better than the competition. But make it clear that while you're willing to pay more than the going rate, you require first rate performance. I've found that paying a decent wage is the best way to guarantee professional quality work.

Besides offering good pay, keep up morale by showing appreciation for a job well done. Be quick to offer verbal praise when it's due. And keep your criticism in bounds. If you have to criticize, do it in private. And always criticize the act, not the person.

When possible, delegate more and more responsibility for on-site decision-making to your crew. This shows that you respect and trust their judgment. When an employee makes an occasional mistake — and don't we all — back them up. Don't revoke your grant of responsibility for a single mistake. Give your employees the same opportunity to learn from their mistakes that you have given to yourself.

Everyone on your payroll should feel like they're an important part of the team, with enough authority to discharge their responsibilities. No one wants to feel like a drone, doing only what they're told to do. And you don't want to have to do all that telling. Encourage your people to develop good judgment and take responsibility for their work. Your goal, and mine in writing this book, is to make this work an enjoyable and enriching experience for yourself, your crew, and those you work for.

Some photographs, tools and equipment were furnished by the following manufacturers and suppliers.

American Design and Engineering, Inc.
900 3rd St.
St. Paul Park, MN 55071
(800) 441-1388

Bosch Power Tool Corporation
100 Bosch Blvd.
New Bern, NC 28562-4097
(800) 334-4151

By George Enterprises, Inc.
P.O. Box 757
Quakertown, PA 18951
(800) 545-9821

Eldenwood Enterprises
50 Tiburon St. #21
San Rafael, CA 94901
(800) 735-1801

GIL-LIFT
1605 North River Blvd.
Independence, MO 64050
(816) 833-0611

Hitachi Power Tools
4487-E Park Dr.
Norcross, GA 30093
(800) 448-2244

Iron Age Safety Shoes
2406 Woodmere Dr.
Pittsburgh, PA 15205
(800) 443-2181

Journeyman Products, Ltd.
P.O. Box 4472
Annapolis, MD 21403
(800) 248-8707

KNAACK Manufacturing Co.
420 East Terra Cotta Ave.
Crystal Lake, IL 60014
(815) 459-6020

Maizefield Mantels Co.
P.O. Box 336
Port Townsend, WA 98368
(206) 385-6789

Makita U.S.A., Inc.
14930 Northam St.
La Mirada, CA 90638-5753
(714) 522-8088

Matrix Enterprises, Inc.
5926 Sedgwick Rd.
West Worthington, OH 43235
(614) 846-0030

Occidental Leather
P.O. Box 364
Valley Ford, CA 94972
(707) 874-3650

Penzotti Pete's
P.O. Box 151023
San Rafael, CA 94901
(415) 456-4059

Price Brothers Tools
P.O. Box 1133
Novato, CA 94948
(800) 334-8270

Rousseau Co.
1712 Thirteenth St.
Clarkston, WA 99403
(509) 758-3954

Senco Products
8485 Broadwell Rd.
Cincinnati, OH 45244
(513) 388-2000

Wedge Innovations
2040 Fortune Dr., Suite 102
San Jose, CA 95131
(408) 434-7000

Wood Dynamics
15034 N.E. 172nd Ave.
Brush Prairie, WA 98606
(206) 896-9047

Woodstock International, Inc.
P.O. Box 2027
Bellingham, WA 98227
(206) 734-3482

Woodworker's Supply of New Mexico
5604 Alameda Place, N.E.
Albuquerque, NM 87113
(800) 645-9292

Index

Practical References for Builders

Working Alone

This unique book shows you how to become a dynamic one-man team as you handle nearly every aspect of house construction, including foundation layout, setting up scaffolding, framing floors, building and erecting walls, squaring up walls, installing sheathing, laying out rafters, raising the ridge, getting the roof square, installing rafters, subfascia, sheathing, finishing eaves, installing windows, hanging drywall, measuring trim, installing cabinets, and building decks. **152 pages, 5¹/₂ x 8¹/₂, $17.95**

Roof Framing

Shows how to frame any type of roof in common use today, even if you've never framed a roof before. Includes using a pocket calculator to figure any common, hip, valley, or jack rafter length in seconds. Over 400 illustrations cover every measurement and every cut on each type of roof: gable, hip, Dutch, Tudor, gambrel, shed, gazebo, and more. **480 pages, 5¹/₂ x 8¹/₂, $22.00**

Rough Framing Carpentry

If you'd like to make good money working outdoors as a framer, this is the book for you. Here you'll find shortcuts to laying out studs; speed cutting blocks, trimmers and plates by eye; quickly building and blocking rake walls; installing ceiling backing, ceiling joists, and truss joists; cutting and assembling hip trusses and California fills; arches and drop ceilings — all with production line procedures that save you time and help you make more money. Over 100 on-the-job photos of how to do it right and what can go wrong. **304 pages, 8¹/₂ x 11, $26.50**

Markup & Profit: A Contractor's Guide

In order to succeed in a construction business, you have to be able to price your jobs to cover all labor, material and overhead expenses, and make a decent profit. The problem is knowing what markup to use. You don't want to lose jobs because you charge too much, and you don't want to work for free because you've charged too little. If you know how to calculate markup, you can apply it to your job costs to find the right sales price for your work. This book gives you tried and tested formulas, with step-by-step instructions and easy-to-follow examples, so you can easily figure the markup that's right for your business. Includes a CD-ROM with forms and checklists for your use. **320 pages, 8¹/₂ x 11, $32.50**

Finish Carpentry

The time-saving methods and proven shortcuts you need to do first class finish work on any job: cornices and rakes, gutters and downspouts, wood shingle roofing, asphalt, asbestos and built-up roofing, prefabricated windows, door bucks and frames, door trim, siding, wallboard, lath and plaster, stairs and railings, cabinets, joinery, and wood flooring. **192 pages, 8¹/₂ x 11, $15.25**

Contractor's Guide to the Building Code Revised

This new edition was written in collaboration with the International Conference of Building Officials, writers of the code. It explains in plain English exactly what the latest edition of the *Uniform Building Code* requires. Based on the 1997 code, it explains the changes and what they mean for the builder. Also covers the *Uniform Mechanical Code* and the *Uniform Plumbing Code*. Shows how to design and construct residential and light commercial buildings that'll pass inspection the first time. Suggests how to work with an inspector to minimize construction costs, what common building shortcuts are likely to be cited, and where exceptions may be granted. **320 pages, 8¹/₂ x 11, $39.00**

Stair Builders Handbook

If you know the floor-to-floor rise, this handbook gives you everything else: number and dimension of treads and risers, total run, correct well hole opening, angle of incline, and quantity of materials and settings for your framing square for over 3,500 code-approved rise and run combinations — several for every 1/8-inch interval from a 3 foot to a 12 foot floor-to-floor rise. **416 pages, 5¹/₂ x 8¹/₂, $19.50**

Moving to Commercial Construction

In commercial work, a single job can keep you and your crews busy for a year or more. The profit percentages are higher, but so is the risk involved. This book takes you step-by-step through the process of setting up a successful commercial business; finding work, estimating and bidding, value engineering, getting through the submittal and shop drawing process, keeping a stable work force, controlling costs, and promoting your business. Explains the design/build and partnering business concepts and their advantage over the competitive bid process. Includes sample letters, contracts, checklists and forms that you can use in your business, plus a CD-ROM with blank copies in several word-processing formats for both Mac and PC computers. **256 pages, 8¹/₂ x 11, $42.00**

National Construction Estimator

Current building costs for residential, commercial, and industrial construction. Estimated prices for every common building material. Provides manhours, recommended crew, and gives the labor cost for installation. Includes a CD-ROM with an electronic version of the book with *National Estimator*, a stand-alone *Windows*™ estimating program, plus an interactive multimedia video that shows how to use the disk to compile construction cost estimates. **656 pages, 8¹/₂ x 11, $47.50. Revised annually**

Contractor's Guide to QuickBooks Pro 2003

This user-friendly manual walks you through QuickBooks Pro's detailed setup procedure and explains step-by-step how to create a first-rate accounting system. You'll learn in days, rather than weeks, how to use QuickBooks Pro to get your contracting business organized, with simple, fast accounting procedures. On the CD included with the book you'll find a QuickBooks Pro file preconfigured for a construction company (you drag it over onto your computer and plug in your own company's data). You'll also get a complete estimating program, including a database, and a job costing program that lets you export your estimates to QuickBooks Pro. It even includes many useful construction forms to use in your business. **336 pages, 8¹/₂ x 11, $47.75**
  Also available: **Contractor's Guide to QuickBooks Pro 2001, $45.25**
  Contractor's Guide to QuickBooks Pro 1999, $42.00

Construction Forms & Contracts

125 forms you can copy and use — or load into your computer (from the FREE disk enclosed). Then you can customize the forms to fit your company, fill them out, and print. Loads into *Word* for *Windows*™, *Lotus 1-2-3*, *WordPerfect, Works*, or *Excel* programs. You'll find forms covering accounting, estimating, fieldwork, contracts, and general office. Each form comes with complete instructions on when to use it and how to fill it out. These forms were designed, tested and used by contractors, and will help keep your business organized, profitable and out of legal, accounting and collection troubles. Includes a CD-ROM for *Windows*™ and Mac. **432 pages, 8¹/₂ x 11, $41.75**

CD Estimator

If your computer has *Windows*™ and a CD-ROM drive, CD Estimator puts at your fingertips 85,000 construction costs for new construction, remodeling, renovation & insurance repair, electrical, plumbing, HVAC and painting. Quarterly cost updates are available at no charge on the Internet. You'll also have the *National Estimator* program — a stand-alone estimating program for *Windows*™ that *Remodeling* magazine called a "computer wiz," and Job Cost Wizard, a program that lets you export your estimates to QuickBooks Pro for actual job costing. A 60-minute interactive video teaches you how to use this CD-ROM to estimate construction costs. And to top it off, to help you create professional-looking estimates, the disk includes over 40 construction estimating and bidding forms in a format that's perfect for nearly any *Windows*™ word processing or spreadsheet program. **CD Estimator is $68.50**

A Roof Cutter's Secrets to Custom Homes

A master framer spills his secrets to framing irregular roofs, jobsite solutions for rake walls, and curved and two-story walls. You'll also find step-by-step techniques for cutting bay roofs, gambrels, and shed, gable, and eyebrow dormers. You'll even find instructions on custom work like coffered ceilings, arches and barrel vaults; even round towers, hexagons, and other polygons. Includes instructions for figuring most of the equations in this book with the keypad of the Construction Master Pro calculator. **342 pages, 8¹/₂ x 5¹/₂, $32.50**

Contractor's Plain-English Legal Guide

For today's contractors, legal problems are like snakes in the swamp - you might not see them, but you know they're there. This book tells you where the snakes are hiding and directs you to the safe path. With the directions in this easy-to-read handbook you're less likely to need a $200-an-hour lawyer. Includes simple directions for starting your business, writing contracts that cover just about any eventuality, collecting what's owed you, filing liens, protecting yourself from unethical subcontractors, and more. For about the price of 15 minutes in a lawyer's office, you'll have a guide that will make many of those visits unnecessary. Includes a CD-ROM with blank copies of all the forms and contracts in the book. **272 pages, 8¹/₂ x 11, $49.50**

How to Succeed With Your Own Construction Business

Everything you need to start your own construction business: setting up the paperwork, finding the work, advertising, using contracts, dealing with lenders, estimating, scheduling, finding and keeping good employees, keeping the books, and coping with success. If you're considering starting your own construction business, all the knowledge, tips, and blank forms you need are here. **336 pages, 8¹/₂ x 11, $28.50**

Finish Carpentry: Efficient Techniques for Custom Interiors

Professional finish carpentry demands expert skills, precise tools, and a solid understanding of how to do the work. This new book explains how to install moldings, paneled walls and ceilings, and just about every aspect of interior trim — including doors and windows. Covers built-in bookshelves, coffered ceilings, and skylight wells and soffits, including paneled ceilings with decorative beams. **288 pages, 8¹/₂ x 11, $34.95**

Contractor's Index to the 1997 *Uniform Building Code*

Finally, there's a common-sense index that helps you quickly and easily find the section you're looking for in the *UBC*. It lists topics under the names builders actually use in construction. Best of all, it gives the full section number and the actual page in the *UBC* where you'll find it. If you need to know the requirements for windows in exit access corridor walls, just look under *Windows*™. You'll find the requirements you need are in Section 1004.3.4.3.2.2 in the *UBC* — on page 115. This practical index was written by a former builder and building inspector who knows the *UBC* from both perspectives. If you hate to spend valuable time hunting through pages of fine print for the information you need, this is the book for you. **192 pages, 8¹/₂ x 11, Loose-leaf, $29.00**

JLC Field Guide to Residential Construction

The ultimate visual quick-reference guide for construction professionals. Over 400 precisely-detailed drawings with clear concise notes and explanations that show you everything from estimating and selecting lumber to foundations, roofing, siding and exteriors. Explains code requirements for all U.S. building codes. **386 pages, 8½ x 5½, $69.95**